Reference Tables for Structural Engineers

Verified Hand-Calculations for Rapid Design

Third edition 2026

www.saadtantawi.com

DEDICATION

In memory of my father, Dr. Hasan Tantawi, PhD (1944-2024)

He was a professor of civil engineering by trade, but to me, he was the man who taught me everything I needed to know about this profession and about life. He was my first teacher and my greatest mentor who dedicated his life to the pursuit of structural integrity and the education of others. I hope these pages honor the standards of excellence and the love for engineering he instilled in me. I wish he could have seen the finished work.

PREFACE

The primary objective of this book is to give structural engineers and design professionals, a comprehensive library of design tables based on rigorous hand calculations that i have developed and utilized throughout my career.

This book is the result of years of structural engineering practice. Like many of you, I spent the early years of my career performing the same hand calculations repeatedly. Eventually, I began cataloging these calculations, turning repetitive math into streamlined tables. These pages are not just data, they are the crystallized output of years of structural design check.

This book covers:

- **Concrete**: Capacities for slabs on grade, elevated slabs, isolated footings, piers and specialized mat foundations.
- **Masonry & Steel:** Quick references for masonry walls and light gauge steel beams.
- **Timber:** Load tables for wood beams and posts.
- **Detailing:** A collection of standard CAD details to assist in drafting
- **Retaining Walls:** A quick calculation on how to design a retaining wall and its foundation.

I would like to add, that the inclusion of **Waste Water Treatment Plant mat foundations** and **elevated slab capacities** offers data that is often difficult to find in standard references.

Whether you are on a job site needing a quick capacity check, or in the office needing to size a member before opening your modeling software, these provide immediate answers based on first principles.

My hope is that this book serves as a trusted companion on your desk, saving you time, reducing calculation fatigue, and allowing you to focus more on the art of structural engineering. As always, any additional comments or inputs are always appreciated.

Saad H. Tantawi (PE)

ABOUT THE AUTHOR

Saad H. Tantawi, PE, is a licensed Professional Engineer with years of experience in structural design and analysis. He holds a Bachelor of Science in Civil Engineering, and a Master of Science degree in Civil Engineering with an emphasis in Structural Designs (2013) both from the University of Tennessee at Knoxville.

With active licensure in multiple-states, Saad H. Tantawi has spent his career bridging the gap between complex theoretical analysis and practical, field-ready solutions. His expertise spans residential, commercial, industrial, and specialized civil infrastructure. The collection of design tables is the culmination of ten years of rigorous manual (hand) calculations, developed to provide fellow engineers with the same speed and reliability he applies to his own professional practice. When not designing complex structures, he is dedicated to mentoring the next generation of engineers in the importance of first-principle calculations.

CONTENTS

CONTENTS

CONTENTS

CONTENTS

CONTENTS

CONTENTS

CONTENTS

CONTENTS

CONTENTS

REINFORCED CONCRETE DESIGN EQUATIONS

COMMON FORMULAS USED IN REINFORCED CONCRETE DESIGN

MODULUS OF ELASTICITY (E_c) = $W_C^{1.5} 33\sqrt{f'_c}$

WHERE,

W_c = UNIT WEIGHT OF CONCRETE (PCF)

f'_c = COMPRESSIVE STRENGTH OF CONCRETE (PSI)

MODULUS OF RUPTURE (f_r) = $7.5\sqrt{f'_c}$

NOTE: MODULUS OF RUPTURE IS THE STRESS AT WHICH FLEXURAL CRACKING WOULD BE EXPECTED TO OCCUR.

CRACKING MOMENT (M_{cr}) = $f_r I / c$

WHERE,

Mcr = CRACKING MOMENT OR THE MOMENT THAT PRODUCES A TENSILE STRESS = TO THE MODULUS OF RUPTURE

f_r = MODULUS OF RUPTURE

I = MOMENT OF INERTIA OF THE CROSS SECTION ABOUT THE NEUTRAL AXIS

c = DISTANCE FROM THE NEUTRAL AXIS TO THE OUTSIDE TENSION OR COMPRESSION FIBER OF THE BEAM

AXIAL STRENGTH ($P_{n,max}$) = $0.80(P_O)$ = $0.80*(0.85f'_c\,(A_g - A_{st}) + f_y A_{st})$

WHERE,

$P_{n,max}$ = MAXIMUM NOMINAL AXIAL COMPRESSIVE STRENGTH AS IS DEFINED BY THE ACI 318R SECTION 22.4.2.1

P_O = NOMINAL AXIAL STRENGTH AT ZERO ECCENTRICITY, LBS.

A_g = GROSS AREA OF CONCRETE SECTION, IN^2.

A_{st} = TOTAL AREA OF LONGITUDINAL REINFORCEMENT INCLUDING BARS OR STEEL SHAPES, AND EXCLUDING PRESTRESSING RINFORCEMENT, IN^2.

EQUIVALENT STRESS DISTRIBUTION FACTOR (β_1)

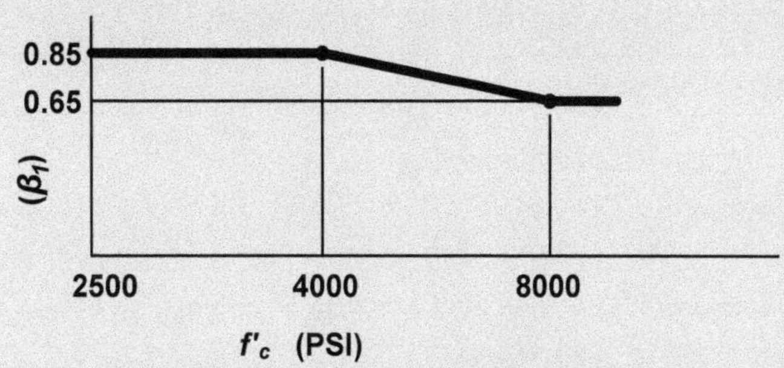

COMMON FORMULAS USED IN REINFORCED CONCRETE DESIGN

LRFD REDUCTION FACTORS (Φ):

TENSION-CONTROLLED SECTIONS:	0.65-0.90 (SEE EQUATION BELOW)
COMPRESSION-CONTROLLED SECTIONS:	0.65 (0.75 FOR SPIRALLY REINFORCED)
SHEAR AND TORSION:	0.75
BEARING ON CONCRETE:	0.65

$\Phi = 0.65 + [(0.90-0.65) / (0.005 - \varepsilon_y)](\varepsilon_t - \varepsilon_y)$

ONE WAY SLAB SHEAR $(\Phi V_N) = 0.75 * \sqrt{f'_c} * bw * d$

WHERE,

bw = WEB WIDTH
d = DISTANCE FROM OUTER COMPRESSIVE FIBER TO TENSILE STEEL REBARS CENTER.

TENSION SPLICES:

LAP SPLICES SHOULD BE CLASS B (LAP LENGTH = 1.3 L_d)

CLASS A LAP SPLICES (LAP LENGTH = 1.0 L_d) ARE ONLY ALLOWED IF:

1- AREA OF STEEL PROVIDED >= 2 * AREA OF STEEL REQUIRED THROUGH THE ENTIRE LENGTH OF THE SPLICE.

2- NOT MORE THAN 50% OF THE TOTAL REINFORCEMENT IS SPLICED WITHIN THE REQUIRED LENGTH.

CRACK CONTROL (TENSION ZONE):

$S = 15\,(40{,}000/f_s) - 2.5C_C \leq 12(40{,}000/f_s)$

WHERE,

S = ENTER-TO-CENTER SPACING OF FLEXURAL TENSION REINFORCEMENT NEAREST TO THE EXTREME TENSION FACE, (INCHES).

f_s = CALCULATED STRESS, (PSI). THIS MAY BE TAKEN AS (2/3) OF THE SPECIFIED YIELD STRENGTH

C_C = CLEAR COVER FROM NEAREST SURFACE IN TENSION TO THE SURFACE OF THE FLEXURAL REINFORCEMENT (INCHES).

EFFECTIVE MOMENT OF INERTIA FOR SLABS:

$0.25\, I_g \leq (0.10 + 25p)(1.2 - 0.2(b_w/d)) * I_g = I_e \leq 0.5 I_g$

WHERE,

p = A_s (STEEL) / A_g

I_g = GROSS MOMENT OF INERTIA = $(b_w * h^3)/12$ FOR A RECTANGULAR SECTION

COMMON REBAR SIZES CHART			
REBAR NUMBER	WEIGHT (LBS/FT)	REBAR DIAMETER (INCHES)	CROSS-SECTIONAL AREA (IN^2)
#3	0.376	0.375	0.11
#4	0.668	0.500	0.20
#5	1.043	0.625	0.31
#6	1.502	0.750	0.44
#7	2.044	0.875	0.60
#8	2.670	1.000	0.79
#9	3.400	1.128	1.00
#10	4.303	1.270	1.27
#11	5.313	1.410	1.56
#14	7.650	1.693	2.25
#18	13.600	2.257	4.00

WELDED WIRE REINFORCEMENT CHART		
NEW DESIGNATION	OLD DESIGNATION	STEEL AREA (IN^2 /FT)
6x6 - W1.4xW1.4	6x6 - 10x10	0.028
6x6 - W2xW2	6x6 - 8x8	0.040
6x6 - W2.9xW2.9	6x6 - 8x8	0.058
6x6 - W4xW4	6x6 - 4x4	0.080
4x4 - W1.4xW1.4	4x4 - 10x10	0.042
4x4 - W2xW2	4x4 - 8x8	0.060

RETAINING WALLS QUICK DESIGN REFERENCE

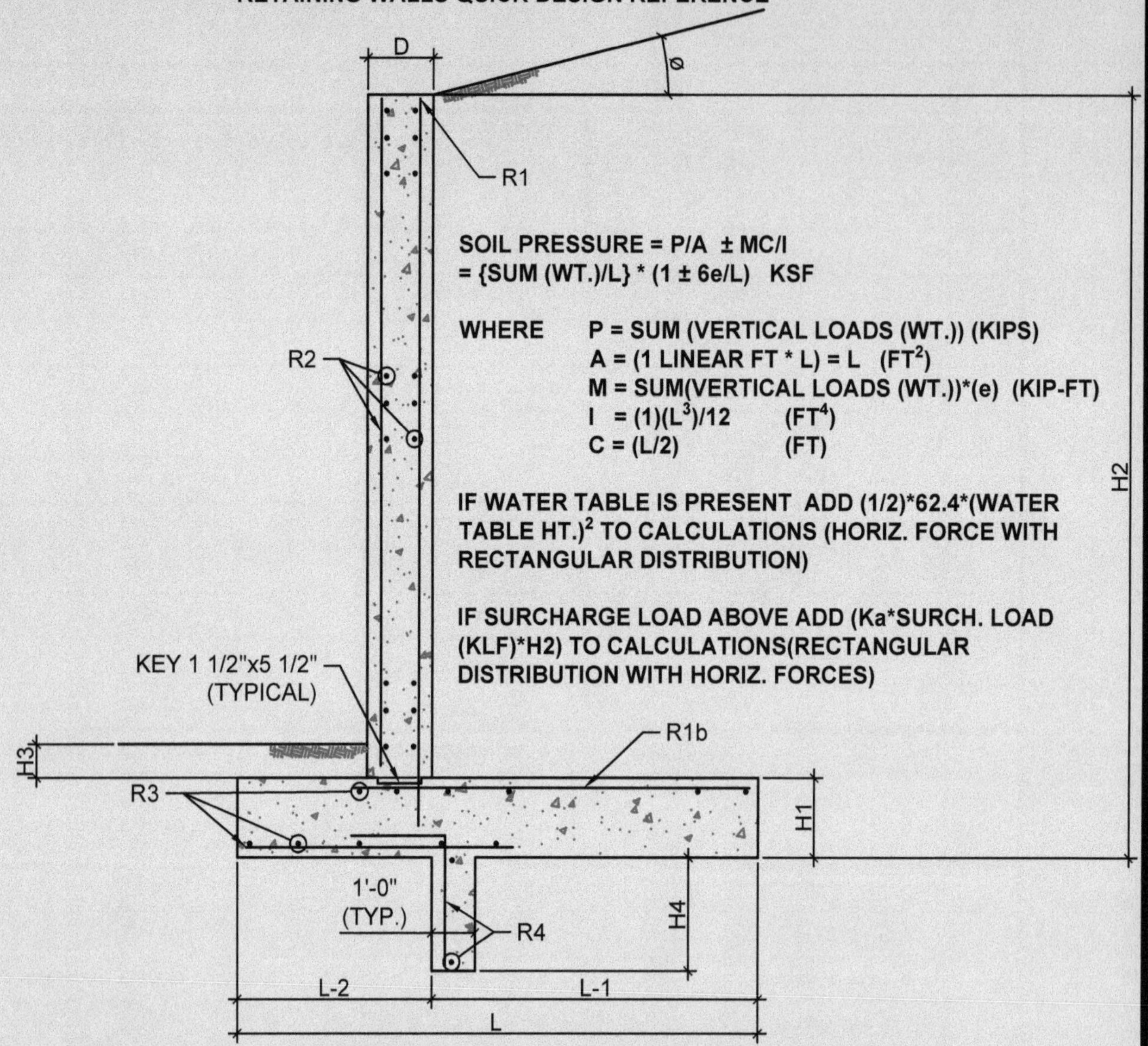

SOIL PRESSURE = P/A ± MC/I
= {SUM (WT.)/L} * (1 ± 6e/L) KSF

WHERE P = SUM (VERTICAL LOADS (WT.)) (KIPS)
A = (1 LINEAR FT * L) = L (FT^2)
M = SUM(VERTICAL LOADS (WT.))*(e) (KIP-FT)
I = (1)(L^3)/12 (FT^4)
C = (L/2) (FT)

IF WATER TABLE IS PRESENT ADD (1/2)*62.4*(WATER TABLE HT.)^2 TO CALCULATIONS (HORIZ. FORCE WITH RECTANGULAR DISTRIBUTION)

IF SURCHARGE LOAD ABOVE ADD (Ka*SURCH. LOAD (KLF)*H2) TO CALCULATIONS(RECTANGULAR DISTRIBUTION WITH HORIZ. FORCES)

"QUICK AND DIRTY" GENERALIZED RETAINING WALLS DESIGN GUIDE

L = (2/3)* H2	e = {(STABILIZING M - OVERTURNING M)/ (SUM OF WT.) } < L/6
H1 = (1/10)*H2	FS (OVERTURNING) > 2.0
L1 = (2/3)*L	FS (SLIDING) > 1.5
L-2 = (1/3)*L	Kp = 1/Ka
H4 = 2'-6"	Ka = (1- SIN (Ω)) / (1 + SIN(Ω))
H3 = VARIES	Y_{SOIL} = WEIGHT OF SOIL (KCF) ---> NOTE: (KCF = 1000 PCF)
D = (1/12)*H2	

Ω = ANGLE OF INTERNAL FRICTION

R1 = AS = FLEXURAL REINFORCEMENT AREA IN STEM =

{(Y_{SOIL}) * (H2^2)*0.5 *(2/3)*H2*Ka*12 / [(0.8*60 KSI*(D)] } < 0.002*D*12 (IN^2/FT)

R2 = 0.002*12*D (IN^2/FT) R3 = 0.002*12*H1 (IN^2/FT) R4 = 0.002*12*12 (IN^2/FT)

R1b = SAME CALCULATIONS AS R1 BUT SUBSTITUTE H1 FOR D

ø < Ω = ANGLE OF INTERNAL FRICTION (SOIL ON SOIL)

SLAB-ON-GRADE CAPACITY (SHEAR, MOMENT, A_{SMIN}) AT VARIOUS DEPTHS

TABLE 2-1

SLAB-ON-GRADE CAPACITY [1,2,3]			
CONCRETE COMPRESSIVE STRENGTH (f'c) = 3000 PSI, MODULUS OF ELASTICITY (E) = 3156 KSI			
SLAB DEPTH (INCHES)	ALLOWABLE ONE WAY SHEAR CAPACITY $(\Phi V_N / \Omega)$ [4,5] (KIPS / LINEAR FT)	ALLOWABLE MOMENT CAPACITY $(\Phi M_N / \Omega)$ [4,5] (KIP-FT / LINEAR FT)	MINIMUM TEMPERATURE AND SHRINKAGE STEEL REQUIRED (A_{SMIN}) [6] (IN^2 / LINEAR FT)
4	0.55	0.44	0.09
5	0.69	0.69	0.11
6	0.83	1.00	0.13
7	0.97	1.36	0.16
8	1.11	1.78	0.18
9	1.25	2.25	0.20
10	1.39	2.78	0.22
11	1.53	3.36	0.24
12	1.67	4.00	0.26
13	1.81	4.70	0.29
14	1.95	5.45	0.31
15	2.09	6.25	0.33
16	2.23	7.12	0.35
17	2.37	8.03	0.37
18	2.51	9.01	0.39

NOTE (1): SLAB-ON-GRADE MINIMUM COVER = 2 INCHES.

NOTE (2): STEEL SHALL BE LOCATED IN THE TOP 2 INCHES, OR THE UPPER 1/3 OF THE SLAB FOR SLAB THICKNESS $\leq$ 8 INCHES. USE TWO LAYERS OF STEEL (ONE IN TOP AND ONE IN BOTTOM) FOR SLAB THICKNESS > 8 INCHES.

NOTE (3) : ASSUME NORMAL WEIGHT CONCRETE (λ = 1), DENSITY = 145 PCF.

NOTE (4): SLAB OWN WEIGHT HAS ALREADY BEEN FACTORED INTO THE CALCULATIONS.

NOTE (5): LRFD FACTOR (Φ) = 0.75 FOR SHEAR, AND 0.65 FOR FLEXURAL CAPACITY. Ω = ASD FACTOR = 1.6

NOTE (6): STEEL REINFORCEMENT IS GRADE 60 STEEL (ASTM A615)

TABLE 2-2

SLAB-ON-GRADE CAPACITY [1,2,3]			
CONCRETE COMPRESSIVE STRENGTH (f'c) = 3500 PSI, MODULUS OF ELASTICITY (E) = 3409 KSI			
SLAB DEPTH (INCHES)	ALLOWABLE ONE WAY SHEAR CAPACITY ($\Phi VN / \Omega$) [4,5] (KIPS / LINEAR FT)	ALLOWABLE MOMENT CAPACITY ($\Phi M_N / \Omega$) [4,5] (KIP-FT / LINEAR FT)	MINIMUM TEMPERATURE AND SHRINKAGE STEEL REQUIRED (A_{SMIN}) (IN^2 / LINEAR FT)
4	0.59	0.47	0.09
5	0.74	0.74	0.11
6	0.89	1.08	0.13
7	1.04	1.46	0.16
8	1.19	1.92	0.18
9	1.35	2.43	0.20
10	1.50	3.00	0.22
11	1.65	3.62	0.24
12	1.80	4.32	0.26
13	1.95	5.07	0.29
14	2.10	5.88	0.31
15	2.25	6.75	0.33
16	2.40	7.69	0.35
17	2.55	8.67	0.37
18	2.71	9.73	0.39

NOTE (1): SLAB-ON-GRADE MINIMUM COVER = 2 INCHES.

NOTE (2): STEEL SHALL BE LOCATED IN THE TOP 2 INCHES, OR THE UPPER 1/3 OF THE SLAB FOR SLAB THICKNESS $\leq$ 8 INCHES.
USE TWO LAYERS OF STEEL (ONE IN TOP AND ONE IN BOTTOM) FOR SLAB THICKNESS > 8 INCHES.

NOTE (3) : ASSUME NORMAL WEIGHT CONCRETE (λ = 1), DENSITY = 145 PCF.

NOTE (4): SLAB OWN WEIGHT HAS ALREADY BEEN FACTORED INTO THE CALCULATIONS.

NOTE (5): LRFD FACTOR (Φ) = 0.75 FOR SHEAR, AND 0.65 FOR FLEXURAL CAPACITY. Ω = ASD FACTOR = 1.6

NOTE (6): STEEL REINFORCEMENT IS GRADE 60 STEEL (ASTM A615)

TABLE 2-3

SLAB-ON-GRADE CAPACITY [1,2,3]			
CONCRETE COMPRESSIVE STRENGTH (f'c) = 4000 PSI, MODULUS OF ELASTICITY (E) = 3644 KSI			
SLAB DEPTH (INCHES)	ALLOWABLE ONE WAY SHEAR CAPACITY ($\Phi VN / \Omega$) [4,5] (KIPS / LINEAR FT)	ALLOWABLE MOMENT CAPACITY ($\Phi M_N / \Omega$) [4,5] (KIP-FT / LINEAR FT)	MINIMUM TEMPERATURE AND SHRINKAGE STEEL REQUIRED (A_{SMIN}) (IN^2 / LINEAR FT)
4	0.63	0.50	0.09
5	0.79	0.79	0.11
6	0.95	1.15	0.13
7	1.11	1.56	0.16
8	1.27	2.05	0.18
9	1.44	2.59	0.20
10	1.60	3.20	0.22
11	1.76	3.86	0.24
12	1.92	4.61	0.26
13	2.08	5.42	0.29
14	2.24	6.28	0.31
15	2.40	7.21	0.33
16	2.56	8.22	0.35
17	2.72	9.26	0.37
18	2.89	10.40	0.39

NOTE (1): SLAB-ON-GRADE MINIMUM COVER = 2 INCHES.

NOTE (2): STEEL SHALL BE LOCATED IN THE TOP 2 INCHES, OR THE UPPER 1/3 OF THE SLAB FOR SLAB THICKNESS $\le$ 8 INCHES. USE TWO LAYERS OF STEEL (ONE IN TOP AND ONE IN BOTTOM) FOR SLAB THICKNESS > 8 INCHES.

NOTE (3) : ASSUME NORMAL WEIGHT CONCRETE (λ = 1), DENSITY = 145 PCF.

NOTE (4): SLAB OWN WEIGHT HAS ALREADY BEEN FACTORED INTO THE CALCULATIONS.

NOTE (5): LRFD FACTOR (Φ) = 0.75 FOR SHEAR, AND 0.65 FOR FLEXURAL CAPACITY. Ω = ASD FACTOR = 1.6

NOTE (6): STEEL REINFORCEMENT IS GRADE 60 STEEL (ASTM A615)

TABLE 2-4

SLAB-ON-GRADE TOTAL ALLOWABLE LOADS BASED ON ACI224R-01 TABLE 4.1: " GUIDE TO REASONABLE CRACK WIDTHS, REINFORCED CONCRETE UNDER SERVICE LOADS" (1)					
TOTAL SERVICE LOADS (PSF) (2)					
SLAB DEPTH (INCHES)	CONCRETE COMPRESSIVE STRENGTH (f'c) = 3000 PSI	CONCRETE COMPRESSIVE STRENGTH (f'c) = 3500 PSI	CONCRETE COMPRESSIVE STRENGTH (f'c) = 4000 PSI	MINIMUM STEEL REQUIRED (A_{SMIN}) (3),(4) (IN^2 / LF)	ESTIMATED CRACK WIDTH AT TOTAL APPLIED SERVICE LOADS (INCHES)
4	284	307	328	0.09	$\leq$ 0.012
5	447	482	502	0.11	$\leq$ 0.012
6	641	695	742	0.13	$\leq$ 0.012
7	876	946	1009	0.16	$\leq$ 0.012
8	1085	1172	1251	0.18	$\leq$ 0.012

NOTE (1): ASSUME SUBGRADE MODULUS = 120 KCF (70 PCI)

NOTE (2): SLAB WEIGHT HAS ALREADY BEEN FACTORED INTO UNDER TOTAL SERVICE LOADS CALCULATION. (NORMAL WEIGHT CONCRETE (λ = 1), DENSITY = 145 PCF)

NOTE (3): STEEL REINFORCEMENT IS GRADE 60 STEEL (ASTM A615)

NOTE (4): STEEL SHALL BE LOCATED IN THE TOP 2 INCHES, OR THE UPPER 1/3 OF THE SLAB.

TABLE 2-5

WASTE WATER TREATMENT TANK/ WATER RETAINING STRUCTURES MAXIMUM ALLOWED WATER DEPTH FOR A SLAB-ON-GRADE / MAT FOUNDATION
BASED ON ACI224R-01 TABLE 4.1: " GUIDE TO REASONABLE CRACK WIDTHS, REINFORCED CONCRETE UNDER SERVICE LOADS" [1]

MAXIMUM ALLOWED WATER DEPTH [2]
(f'c = 4000 PSI , MAXIMUM CRACK WIDTH ALLOWED AT SERVICE LOADS $\leq$ 0.004 IN)

SLAB DEPTH (INCHES)	MAXIMUM ALLOWED WATER HEIGHT	REINFORCING STEEL REQUIRED BOTTOM LAYER ($A_{S\,(req'd)}$) [3] (IN^2 / LF)	REINFORCING STEEL REQUIRED TOP LAYER ($A_{S'(req'd)}$) [3] (IN^2 / LF)	CONCRETE COVER BOTT. (INCHES)	CONCRETE COVER TOP (INCHES)
8	4'- 10"	#5 @ 10" o.c.	#4 @ 8" o.c.	3	2
9	5'- 4"	#5 @ 9" o.c.	#4 @ 7" o.c.	3	2
10	6'- 0"	#6 @ 10" o.c.	#5 @ 9" o.c.	3	2
11	6'- 8"	#6 @ 8" o.c.	#6 @ 9" o.c.	3	2
12	7'- 0"	#6 @ 8" o.c.	#6 @ 9" o.c.	3	2
12	7'- 9"	#6 @ 6" o.c.	#6 @ 7" o.c.	3	2
14	8'- 8"	#7 @ 7" o.c.	#7 @ 8" o.c.	3	2
16	9'- 4"	#7 @ 6" o.c.	#7 @ 7" o.c.	3	2
18	9'- 11"	#7 @ 6" o.c.	#7 @ 6" o.c.	3	2
20	10'- 3"	#8 @ 8" o.c.	#7 @ 6" o.c.	3	2
21	10'- 5"	#8 @ 8" o.c.	#7 @ 6" o.c.	3	2
24	10'- 11"	#8 @ 8" o.c.	#7 @ 6" o.c.	3	2
24	12'- 0"	#8 @ 6" o.c.	#8 @ 8" o.c.	3	2
28	12'- 8"	#8 @ 6" o.c.	#8 @ 6" o.c.	3	2
30	13'- 1"	#8 @ 6" o.c.	#8 @ 6" o.c.	3	2
36	13'- 9"	#8 @ 6" o.c.	#8 @ 6" o.c.	3	2

NOTE (1): ASSUME SUBGRADE MODULUS = 120 KCF (70 PCI)

NOTE (2): SLAB WEIGHT HAS ALREADY BEEN FACTORED INTO UNDER TOTAL SERVICE LOADS CALCULATIONS. (NORMAL WEIGHT CONCRETE (λ = 1), DENSITY = 145 PCF)

NOTE (3): STEEL REINFORCEMENT IS GRADE 60 STEEL (ASTM A615)

ELEVATED SLAB TOTAL ALLOWABLE LOADS (PSF) AT VARIOUS DEPTHS

TABLE 3-1

ELEVATED SLAB ALLOWABLE SERVICE LOADS
BASED ON ALLOWABLE MOMENT CAPACITY AND / OR ESTIMATED CRACK WIDTH AT SERVICE LOADS TO BE $\leq$ 0.016 IN, WHICHEVER CONTROLS
"SEE ACI224R-01 TABLE 4.1 FOR CRACK WIDTH RECOMMENDATION"

ELEVATED SLAB TOTAL ALLOWABLE LOADS [(1)]

SLAB DEPTH (INCHES)	CLEAR SPAN OF CONCRETE SLAB (L_n) (FT)	MINIMUM TEMP. AND SHRINKAGE STEEL REQ'D (A_{SMIN}) (IN^2/FT) [(2),(4)]	CONCRETE COMPRESSIVE STRENGTH (f'c) (PSI)	TOTAL ALLOWABLE LOAD (PSF) [(3)]	EFFECTIVE MOMENT OF INERTIA PER LINEAR FT OF WIDTH (IN^4)
4	3'- 6"	0.09	3000	231	16
4	4'- 0"	0.09	3000	164	16
4	4'- 6"	0.09	3000	117	16
4	5'- 0"	0.09	3000	84	16
4	5'- 4"	0.09	3000	66	16
4	3'- 6"	0.09	3500	255	16
4	4'- 0"	0.09	3500	182	16
4	4'- 6"	0.09	3500	131	16
4	5'- 0"	0.09	3500	95	16
4	5'- 4"	0.09	3500	76	16
4	3'- 6"	0.09	4000	277	16
4	4'- 0"	0.09	4000	198	16
4	4'- 6"	0.09	4000	144	16
4	5'- 0"	0.09	4000	106	16
4	5'- 4"	0.09	4000	86	16

NOTE (1): SLAB WEIGHT HAS ALREADY BEEN FACTORED INTO UNDER TOTAL ALLOWABLE LOADS CALCULATIONS. (NORMAL WEIGHT CONCRETE (λ = 1), DENSITY = 145 PCF).

NOTE (2): STEEL REINFORCEMENT IS GRADE 60 STEEL (ASTM A615).

NOTE (3): IF MOMENT CAPACITY CONTROLS, FACTOR OF SAFETY USED (FS = 1.6 / 0.65) OR ABOUT 2.5

NOTE (4): REINFORCING STEEL TO BE PLACED IN TOP 2 INCHES, OR THE UPPER 1/3 OF THE SLAB.

TABLE 3-2

ELEVATED SLAB ALLOWABLE SERVICE LOADS
BASED ON ALLOWABLE MOMENT CAPACITY AND / OR ESTIMATED CRACK WIDTH AT SERVICE LOADS TO BE ≤ 0.016 IN, WHICHEVER CONTROLS
"SEE ACI224R-01 TABLE 4.1 FOR CRACK WIDTH RECOMMENDATION"

ELEVATED SLAB TOTAL ALLOWABLE LOADS [1]					
SLAB DEPTH (INCHES)	CLEAR SPAN OF CONCRETE SLAB (L_n) (FT)	MINIMUM TEMP. AND SHRINKAGE STEEL REQ'D (A_{SMIN}) (IN^2/FT) [2],[4]	CONCRETE COMPRESSIVE STRENGTH (f'c) (PSI)	TOTAL ALLOWABLE LOAD (PSF) [3]	EFFECTIVE MOMENT OF INERTIA PER LINEAR FT OF WIDTH (IN^4)
5	4'- 0"	0.11	3000	275	31.25
5	4'- 6"	0.11	3000	202	31.25
5	5'- 0"	0.11	3000	149	31.25
5	5'- 6"	0.11	3000	111	31.25
5	6'- 0"	0.11	3000	81	31.25
5	6'- 8"	0.11	3000	52	31.25
5	4'- 0"	0.11	3500	303	31.25
5	4'- 6"	0.11	3500	224	31.25
5	5'- 0"	0.11	3500	167	31.25
5	5'- 6"	0.11	3500	126	31.25
5	6'- 0"	0.11	3500	94	31.25
5	6'- 8"	0.11	3500	62	31.25
5	4'- 0"	0.11	4000	328	31.25
5	4'- 6"	0.11	4000	244	31.25
5	5'- 0"	0.11	4000	184	31.25
5	5'- 6"	0.11	4000	139	31.25
5	6'- 0"	0.11	4000	105	31.25
5	6'- 8"	0.11	4000	70	31.25

NOTE (1): SLAB WEIGHT HAS ALREADY BEEN FACTORED INTO UNDER TOTAL ALLOWABLE LOADS CALCULATIONS. (NORMAL WEIGHT CONCRETE (λ = 1), DENSITY = 145 PCF).

NOTE (2): STEEL REINFORCEMENT IS GRADE 60 STEEL (ASTM A615).

NOTE (3): IF MOMENT CAPACITY CONTROLS, FACTOR OF SAFETY USED (FS = 1.6 / 0.65) OR ABOUT 2.5

NOTE (4): REINFORCING STEEL TO BE PLACED IN TOP 2 INCHES, OR THE UPPER 1/3 OF THE SLAB.

TABLE 3-3

ELEVATED SLAB ALLOWABLE SERVICE LOADS
BASED ON ALLOWABLE MOMENT CAPACITY AND / OR ESTIMATED CRACK WIDTH AT SERVICE LOADS TO BE ≤ 0.016 IN, WHICHEVER CONTROLS
"SEE ACI224R-01 TABLE 4.1 FOR CRACK WIDTH RECOMMENDATION"

ELEVATED SLAB TOTAL ALLOWABLE LOADS (1)

SLAB DEPTH (INCHES)	CLEAR SPAN OF CONCRETE SLAB (L_n) (FT)	MINIMUM TEMP. AND SHRINKAGE STEEL REQ'D (A_{SMIN}) (IN^2/FT) (2),(4)	CONCRETE COMPRESSIVE STRENGTH (f'c) (PSI)	TOTAL ALLOWABLE LOAD (PSF) (3)	EFFECTIVE MOMENT OF INERTIA PER LINEAR FT OF WIDTH (IN^4)
6	5'- 6"	0.13	3000	177	54
6	6'- 0"	0.13	3000	135	54
6	6'- 6"	0.13	3000	102	54
6	7'- 0"	0.13	3000	76	54
6	7'- 6"	0.13	3000	55	54
6	8'- 0"	0.13	3000	38	54
6	5'- 6"	0.13	3500	198	54
6	6'- 0"	0.13	3500	153	54
6	6'- 6"	0.13	3500	117	54
6	7'- 0"	0.13	3500	89	54
6	7'- 6"	0.13	3500	66	54
6	8'- 0"	0.13	3500	48	54
6	5'- 6"	0.13	4000	217	54
6	6'- 0"	0.13	4000	168	54
6	6'- 6"	0.13	4000	130	54
6	7'- 0"	0.13	4000	100	54
6	7'- 6"	0.13	4000	76	54
6	8'- 0"	0.13	4000	56	54

NOTE (1): SLAB WEIGHT HAS ALREADY BEEN FACTORED INTO UNDER TOTAL ALLOWABLE LOADS CALCULATIONS. (NORMAL WEIGHT CONCRETE (λ = 1), DENSITY = 145 PCF).

NOTE (2): STEEL REINFORCEMENT IS GRADE 60 STEEL (ASTM A615).

NOTE (3): IF MOMENT CAPACITY CONTROLS, FACTOR OF SAFETY USED (FS = 1.6 / 0.65) OR ABOUT 2.5

NOTE (4): REINFORCING STEEL TO BE PLACED IN TOP 2 INCHES, OR THE UPPER 1/3 OF THE SLAB.

TABLE 3-4

ELEVATED SLAB ALLOWABLE SERVICE LOADS BASED ON ALLOWABLE MOMENT CAPACITY AND / OR ESTIMATED CRACK WIDTH AT SERVICE LOADS TO BE ≤ 0.016 IN, WHICHEVER CONTROLS "SEE ACI224R-01 TABLE 4.1 FOR CRACK WIDTH RECOMMENDATION"					
ELEVATED SLAB TOTAL ALLOWABLE LOADS [(1)]					
SLAB DEPTH (INCHES)	CLEAR SPAN OF CONCRETE SLAB (L_n) (FT)	MINIMUM TEMP. AND SHRINKAGE STEEL REQ'D (A_{SMIN}) (IN^2/FT) [(2),(4)]	CONCRETE COMPRESSIVE STRENGTH (f'c) (PSI)	TOTAL ALLOWABLE LOAD (PSF) [(3)]	EFFECTIVE MOMENT OF INERTIA PER LINEAR FT OF WIDTH (IN^4)
7	6'- 6"	0.16	3000	156	85.75
7	7'- 0"	0.16	3000	120	85.75
7	7'- 6"	0.16	3000	91	85.75
7	8'- 0"	0.16	3000	68	85.75
7	8'- 6"	0.16	3000	49	85.75
7	9'- 4"	0.16	3000	23	85.75
7	6'- 6"	0.16	3500	176	85.75
7	7'- 0"	0.16	3500	138	85.75
7	7'- 6"	0.16	3500	107	85.75
7	8'- 0"	0.16	3500	82	85.75
7	8'- 6"	0.16	3500	61	85.75
7	9'- 4"	0.16	3500	33	85.75
7	6'- 6"	0.16	4000	195	85.75
7	7'- 0"	0.16	4000	154	85.75
7	7'- 6"	0.16	4000	121	85.75
7	8'- 0"	0.16	4000	94	85.75
7	8'- 6"	0.16	4000	72	85.75
7	9'- 4"	0.16	4000	42	85.75

NOTE (1): SLAB WEIGHT HAS ALREADY BEEN FACTORED INTO UNDER TOTAL ALLOWABLE LOADS CALCULATIONS. (NORMAL WEIGHT CONCRETE (λ = 1), DENSITY = 145 PCF).

NOTE (2): STEEL REINFORCEMENT IS GRADE 60 STEEL (ASTM A615).

NOTE (3): IF MOMENT CAPACITY CONTROLS, FACTOR OF SAFETY USED (FS = 1.6 / 0.65) OR ABOUT 2.5

NOTE (4): REINFORCING STEEL TO BE PLACED IN TOP 2 INCHES, OR THE UPPER 1/3 OF THE SLAB.

TABLE 3-5

ELEVATED SLAB ALLOWABLE SERVICE LOADS BASED ON ALLOWABLE MOMENT CAPACITY AND / OR ESTIMATED CRACK WIDTH AT SERVICE LOADS TO BE ≤ 0.016 IN, WHICHEVER CONTROLS "SEE ACI224R-01 TABLE 4.1 FOR CRACK WIDTH RECOMMENDATION"					
ELEVATED SLAB TOTAL ALLOWABLE LOADS [1]					
SLAB DEPTH (INCHES)	CLEAR SPAN OF CONCRETE SLAB (L_n) (FT)	MINIMUM TEMP. AND SHRINKAGE STEEL REQ'D (A_{SMIN}) (IN^2/FT) [2],[4]	CONCRETE COMPRESSIVE STRENGTH (f'c) (PSI)	TOTAL ALLOWABLE LOAD (PSF) [3]	EFFECTIVE MOMENT OF INERTIA PER LINEAR FT OF WIDTH (IN^4)
8	8'- 0"	0.18	3000	106	128
8	8'- 6"	0.18	3000	81	128
8	9'- 0"	0.18	3000	59	128
8	9'- 6"	0.18	3000	41	128
8	10'- 0"	0.18	3000	26	128
8	10'- 8"	0.18	3000	9	128
8	8'- 0"	0.18	3500	124	128
8	8'- 6"	0.18	3500	96	128
8	9'- 0"	0.18	3500	73	128
8	9'- 6"	0.18	3500	54	128
8	10'- 0"	0.18	3500	37	128
8	10'- 8"	0.18	3500	19	128
8	8'- 0"	0.18	4000	140	128
8	8'- 6"	0.18	4000	111	128
8	9'- 0"	0.18	4000	86	128
8	9'- 6"	0.18	4000	66	128
8	10'- 0"	0.18	4000	48	128
8	10'- 8"	0.18	4000	28	128

NOTE (1): SLAB WEIGHT HAS ALREADY BEEN FACTORED INTO UNDER TOTAL ALLOWABLE LOADS CALCULATIONS. (NORMAL WEIGHT CONCRETE (λ = 1), DENSITY = 145 PCF).

NOTE (2): STEEL REINFORCEMENT IS GRADE 60 STEEL (ASTM A615).

NOTE (3): IF MOMENT CAPACITY CONTROLS, FACTOR OF SAFETY USED (FS = 1.6 / 0.65) OR ABOUT 2.5

NOTE (4): REINFORCING STEEL TO BE PLACED IN TOP 2 INCHES, OR THE UPPER 1/3 OF THE SLAB.

TABLE 3-6

ELEVATED SLAB ALLOWABLE SERVICE LOADS
BASED ON ALLOWABLE MOMENT CAPACITY AND / OR ESTIMATED CRACK WIDTH AT SERVICE LOADS TO BE $\leq$ 0.016 IN, WHICHEVER CONTROLS
"SEE ACI224R-01 TABLE 4.1 FOR CRACK WIDTH RECOMMENDATION"

ELEVATED SLAB TOTAL ALLOWABLE LOADS (1)

CONCRETE COMPRESSIVE STRENGTH (f'c) = 4000 PSI

CONCRETE COVER FOR TOP STEEL REINFORCEMENT = 2 INCHES
CONCRETE COVER FOR BOTTOM STEEL REINFORCEMENT = 3 INCHES

SLAB DEPTH (INCHES)	CLEAR SPAN OF CONCRETE SLAB (L_n) (FT)	BOTTOM LAYER REQ'D STEEL AREA ($A_{S\,(req'd)}$) (2) (IN^2 / LF)	TOP LAYER REQ'D STEEL AREA ($A_{S'(req'd)}$) (2) (IN^2 / LF)	TOTAL ALLOWABLE LOAD (PSF) (3)	EFFECTIVE MOMENT OF INERTIA PER LINEAR FT OF WIDTH (IN^4)
9	8'- 6"	#5 @ 12" o.c.	#4 @ 10" o.c.	199	182.25
9	9'- 0"	#5 @ 12" o.c.	#4 @ 10" o.c.	166	182.25
9	9'- 6"	#5 @ 12" o.c.	#4 @ 10" o.c.	138	182.25
9	10'- 0"	#5 @ 12" o.c.	#4 @ 10" o.c.	114	182.25
9	10'- 6"	#5 @ 12" o.c.	#4 @ 10" o.c.	93	182.25
9	11'- 0"	#5 @ 12" o.c.	#4 @ 10" o.c.	75	182.25
9	11'- 6"	#5 @ 12" o.c.	#4 @ 10" o.c.	59	182.25
9	12'- 0"	#5 @ 12" o.c.	#4 @ 10" o.c.	45	182.25
10	9'- 6"	#6 @ 12" o.c.	#5 @ 10" o.c.	282	250
10	10'- 0"	#6 @ 12" o.c.	#5 @ 10" o.c.	243	250
10	10'- 6"	#6 @ 12" o.c.	#5 @ 10" o.c.	209	250
10	11'- 0"	#6 @ 12" o.c.	#5 @ 10" o.c.	179	250
10	11'- 6"	#6 @ 12" o.c.	#5 @ 10" o.c.	154	250
10	12'- 0"	#6 @ 12" o.c.	#5 @ 10" o.c.	131	250
10	12'- 6"	#6 @ 12" o.c.	#5 @ 10" o.c.	112	250
10	13'- 0"	#6 @ 12" o.c.	#5 @ 10" o.c.	94	250
10	13'- 4"	#6 @ 12" o.c.	#5 @ 10" o.c.	84	250

NOTE (1): SLAB WEIGHT HAS ALREADY BEEN FACTORED INTO UNDER TOTAL ALLOWABLE LOADS CALCULATIONS. (NORMAL WEIGHT CONCRETE (λ = 1), DENSITY = 145 PCF).

NOTE (2): STEEL REINFORCEMENT IS GRADE 60 STEEL (ASTM A615).

NOTE (3): IF MOMENT CAPACITY CONTROLS, FACTOR OF SAFETY USED (FS = 1.6 / 0.65) OR ABOUT 2.5

TABLE 3-7

ELEVATED SLAB ALLOWABLE SERVICE LOADS
BASED ON ALLOWABLE MOMENT CAPACITY AND / OR ESTIMATED CRACK WIDTH AT SERVICE LOADS TO BE $\leq$ 0.016 IN, WHICHEVER CONTROLS
"SEE ACI224R-01 TABLE 4.1 FOR CRACK WIDTH RECOMMENDATION"

ELEVATED SLAB TOTAL ALLOWABLE LOADS [(1)]

CONCRETE COMPRESSIVE STRENGTH (f'c) = 4000 PSI

CONCRETE COVER FOR TOP STEEL REINFORCEMENT = 2 INCHES
CONCRETE COVER FOR BOTTOM STEEL REINFORCEMENT = 3 INCHES

SLAB DEPTH (INCHES)	CLEAR SPAN OF CONCRETE SLAB (L_n) (FT)	BOTTOM LAYER REQ'D STEEL AREA ($A_{S\,(req'd)}$) [(2)] (IN^2 / LF)	TOP LAYER REQ'D STEEL AREA ($A_{S'(req'd)}$) [(2)] (IN^2 / LF)	TOTAL ALLOWABLE LOAD (PSF) [(3)]	EFFECTIVE MOMENT OF INERTIA PER LINEAR FT OF WIDTH (IN^4)
11	11'- 0"	#6 @ 12" o.c.	#5 @ 10" o.c.	206	332.75
11	11'- 6"	#6 @ 12" o.c.	#5 @ 10" o.c.	177	332.75
11	12'- 0"	#6 @ 12" o.c.	#5 @ 10" o.c.	152	332.75
11	12'- 6"	#6 @ 12" o.c.	#5 @ 10" o.c.	129	332.75
11	13'- 0"	#6 @ 12" o.c.	#5 @ 10" o.c.	109	332.75
11	13'- 6"	#6 @ 12" o.c.	#5 @ 10" o.c.	92	332.75
11	14'- 0"	#6 @ 12" o.c.	#5 @ 10" o.c.	76	332.75
11	14'- 8"	#6 @ 12" o.c.	#5 @ 10" o.c.	57	332.75
12	12'- 0"	#7 @ 12" o.c.	#6 @ 12" o.c.	257	432
12	12'- 6"	#7 @ 12" o.c.	#6 @ 12" o.c.	226	432
12	13'- 0"	#7 @ 12" o.c.	#6 @ 12" o.c.	198	432
12	13'- 6"	#7 @ 12" o.c.	#6 @ 12" o.c.	173	432
12	14'- 0"	#7 @ 12" o.c.	#6 @ 12" o.c.	151	432
12	14'- 6"	#7 @ 12" o.c.	#6 @ 12" o.c.	130	432
12	15'- 0"	#7 @ 12" o.c.	#6 @ 12" o.c.	112	432
12	15'- 6"	#7 @ 12" o.c.	#6 @ 12" o.c.	96	432
12	16'- 0"	#7 @ 12" o.c.	#6 @ 12" o.c.	81	432

NOTE (1): SLAB WEIGHT HAS ALREADY BEEN FACTORED INTO UNDER TOTAL ALLOWABLE LOADS CALCULATIONS. (NORMAL WEIGHT CONCRETE (λ = 1), DENSITY = 145 PCF).

NOTE (2): STEEL REINFORCEMENT IS GRADE 60 STEEL (ASTM A615).

NOTE (3): IF MOMENT CAPACITY CONTROLS, FACTOR OF SAFETY USED (FS = 1.6 / 0.65) OR ABOUT 2.5

ISOLATED FOOTINGS, AND REINFORCED CONCRETE PIERS LOADING CAPACITY

TABLE 4-1

ISOLATED FOOTINGS AXIAL LOADS CAPACITY

REQUIRED ISOLATED FOOTING SIZES BASED ON APPLIED AXIAL LOADS (F.S = 2.5)			
MARK	SIZE	MINIMUM REINFORCEMENT REQUIRED	AXIAL LOAD CAPACITY (KIPS) [2]
f'_c = 3000 PSI, ALLOWABLE SOIL BEARING PRESSURE = 2000 PSF			
F3.5	3'-6"x3'-6"x1'-0"	As (MIN) = 0.91 IN^2 E.W, BOTT.	21 .0
F4.5	4'-6"x4'-6"x1'-0"	As (MIN) = 1.2 IN^2 E.W, BOTT.	35 .0
F5.5	5'-6"x5'-6"x1'-0"	As (MIN) = 1.43 IN^2 E.W, BOTT.	51 .0
F6.5 [1]	6'-6"x6'-6"x1'-2"	As (MIN) = 2.0 IN^2 E.W, BOTT.	72 .0
F7.5 [1]	7'-6"x7'-6"x1'-2"	As (MIN) = 2.3 IN^2 E.W, BOTT.	94 .0
F8.5 [1]	8'-6"x8'-6"x1'-4"	As (MIN) = 2.94 IN^2 E.W, BOTT.	122.0
F9.5 [1]	9'-6"x9'-6"x1'-6"	As (MIN) = 3.70 IN^2 E.W, BOTT.	144.0
f'_c = 3000 PSI, ALLOWABLE SOIL BEARING PRESSURE = 2500 PSF			
F3.5	3'-6"x3'-6"x1'-0"	As (MIN) = 0.91 IN^2 E.W, BOTT.	27 .0
F4.5	4'-6"x4'-6"x1'-0"	As (MIN) = 1.2 IN^2 E.W, BOTT.	45 .0
F5.5	5'-6"x5'-6"x1'-0"	As (MIN) = 1.43 IN^2 E.W, BOTT.	68 .0
F6.5 [1]	6'-6"x6'-6"x1'-2"	As (MIN) = 2.0 IN^2 E.W, BOTT.	94 .0
F7.5 [1]	7'-6"x7'-6"x1'-4"	As (MIN) = 2.6 IN^2 E.W, BOTT.	124 .0
F8.5 [1]	8'-6"x8'-6"x1'-7"	As (MIN) = 3.5 IN^2 E.W, BOTT.	156.0
F9.5 [1]	9'-6"x9'-6"x1'-9"	As (MIN) = 4.3 IN^2 E.W, BOTT.	190.0
f'_c = 3000 PSI, ALLOWABLE SOIL BEARING PRESSURE = 3000 PSF			
F3.5	3'-6"x3'-6"x1'-0"	As (MIN) = 0.91 IN^2 E.W, BOTT.	33.0
F4.5	4'-6"x4'-6"x1'-0"	As (MIN) = 1.2 IN^2 E.W, BOTT.	55 .0
F5.5 [1]	5'-6"x5'-6"x1'-1"	As (MIN) = 1.55 IN^2 E.W, BOTT.	82 .0
F6.5 [1]	6'-6"x6'-6"x1'-3"	As (MIN) = 2.11 IN^2 E.W, BOTT.	114 .0
F7.5 [1]	7'-6"x7'-6"x1'-6"	As (MIN) = 2.92 IN^2 E.W, BOTT.	150 .0
F8.5 [1]	8'-6"x8'-6"x1'-9"	As (MIN) = 3.90 IN^2 E.W, BOTT.	190.0
F9.5 [1]	9'-6"x9'-6"x2'-0"	As (MIN) = 4.93 IN^2 E.W, BOTT.	232.0

NOTE (1) : INDICATES FOOTING DEPTHS INCREASED TO ACCOUNT FOR REQUIRED PUNCHING SHEAR CAPACITY

NOTE (2) : FACTOR OF SAFETY USED = 2.5

TABLE 4-2

36"x36" REINFORCED CONCRETE PIERS					
f'c = 3500 PSI , NORMAL WEIGHT CONC. (WT = 0.145 K/FT^3)					
HEIGHT (FEET)	ALLOWABLE AXIAL LOADS P(1),(2) (M,V,T = 0) (KIPS)	ALLOWABLE MOMENT M (1),(2) (P,V,T = 0) (KIP-FT)	ALLOWABLE SHEAR V (1),(2) (P,M,T = 0) (KIPS)	ALLOWABLE TORSION T (1),(2) (P,M,V = 0) (KIP-FT)	SLENDERNESS RATIO (3),(4) KLu/r (KIP-FT)
32	1109.37	646.24	26.25	26.95	35.55
28	1500.00	646.24	29.37	26.95	31.11
24	1500.00	646.24	33.12	26.95	26.66
20	1500.00	646.24	38.11	26.95	22.22
16	1500.00	646.24	45.61	26.95	17.77
12	1500.00	646.24	58.12	26.95	13.33
10	1500.00	646.24	68.74	26.95	11.11

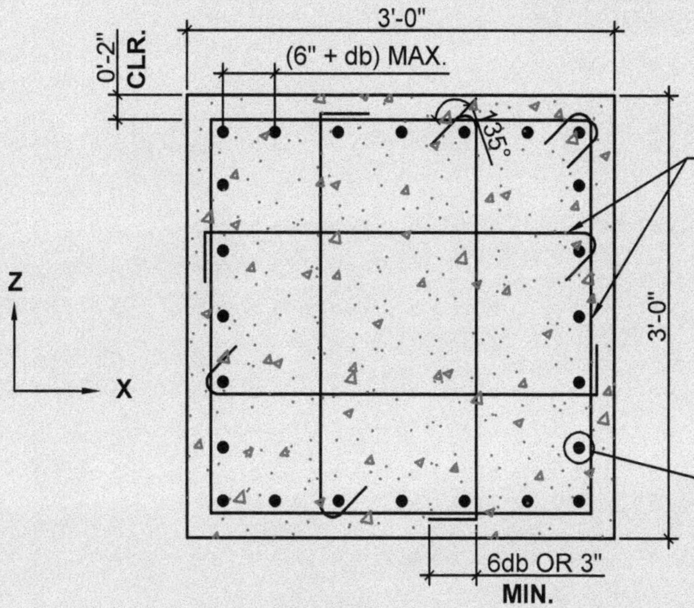

$I_{service}$ = 140107.96 IN^4
E= 3409 KSI

db = TIES DIA.
dbv = VERT. REINF. DIA.

As / Ag = 0.011

NOTE (1) : ALLOWABLE LOADS, MOMENT, SHEAR AND TORSION, ARE TAKEN AS LRFD RESULTS (PHI*Pn, PHI*Mn, PHI*Vn, OR PHI*Tn) / (Ω = 1.6) IN OTHER WORDS TO CONVERT FROM ASD TO LRFD RESULTS MULTIPLY TABULATED RESULTS ABOVE BY A FACTOR OF 1.6

NOTE (2): REINFORCING STEEL IS GR. 60 ASTM A615

NOTE (3): RADIUS OF GYRATION = 0.30*H = 10.8 INCHES, WHERE H = OVERALL DIMENSION OF A RECTANGULAR COLUMN IN THE DIRECTION OF THE MOMENT

NOTE (4): TABULATED RESULTS ARE BASED ON K = EFFECTIVE LENGTH FACTOR FOR COMPRESSION MEMBERS = 1.0

TABLE 4-3

36"x36" REINFORCED CONCRETE PIERS					
f'c = 4000 PSI , NORMAL WEIGHT CONC. (WT = 0.145 K/FT^3)					
HEIGHT (FEET)	ALLOWABLE AXIAL LOADS P[(1),(2)] (M,V,T = 0) (KIPS)	ALLOWABLE MOMENT M[(1),(2)] (P,V,T = 0) (KIP-FT)	ALLOWABLE SHEAR V[(1),(2)] (P,M,T = 0) (KIPS)	ALLOWABLE TORSION T[(1),(2)] (P,M,V = 0) (KIP-FT)	SLENDERNESS RATIO[(3),(4)] KLu/r (KIP-FT)
32	1374.99	650.93	26.25	28.74	35.55
28	1687.5	650.93	29.37	28.74	31.11
24	1687.5	650.93	33.12	28.74	26.66
20	1687.5	650.93	38.11	28.74	22.22
16	1687.5	650.93	45.61	28.74	17.77
12	1687.5	650.93	58.12	28.74	13.33
10	1687.5	650.93	68.74	28.74	11.11

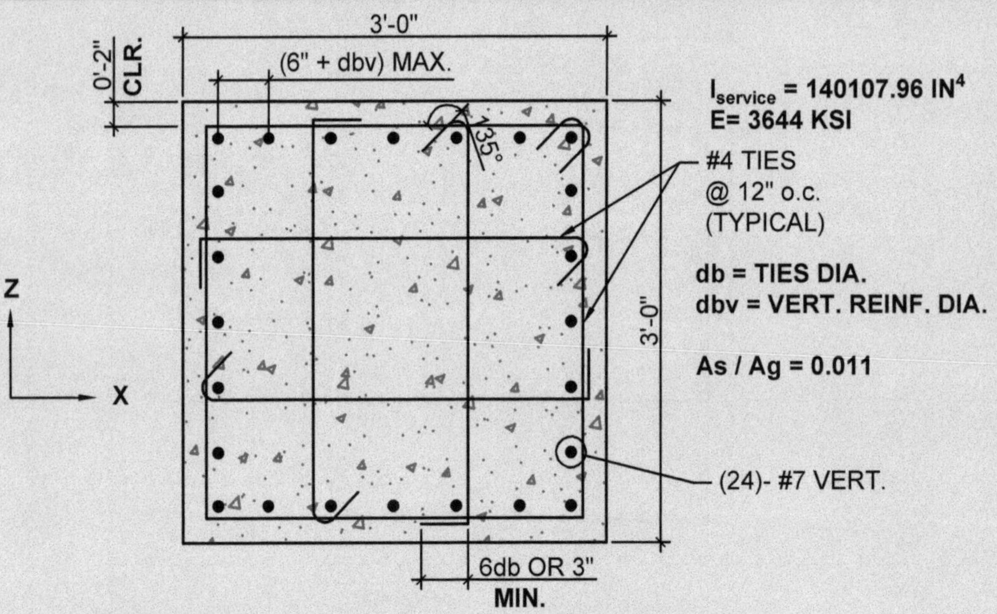

NOTE (1) : ALLOWABLE LOADS, MOMENT, SHEAR AND TORSION, ARE TAKEN AS LRFD RESULTS (PHI*Pn, PHI*Mn, PHI*Vn, OR PHI*Tn) / (Ω = 1.6) IN OTHER WORDS TO CONVERT FROM ASD TO LRFD RESULTS MULTIPLY TABULATED RESULTS ABOVE BY A FACTOR OF 1.6

NOTE (2): REINFORCING STEEL IS GR. 60 ASTM A615

NOTE (3): RADIUS OF GYRATION = 0.30*H = 10.8 INCHES, WHERE H = OVERALL DIMENSION OF A RECTANGULAR COLUMN IN THE DIRECTION OF THE MOMENT

NOTE (4): TABULATED RESULTS ARE BASED ON K = EFFECTIVE LENGTH FACTOR FOR COMPRESSION MEMBERS = 1.0

TABLE 4-4

36"x36" REINFORCED CONCRETE PIERS					
f'c = 4500 PSI , NORMAL WEIGHT CONC. (WT = 0.145 K/FT^3)					
HEIGHT (FEET)	ALLOWABLE AXIAL LOADS $P^{(1),(2)}$ (M,V,T = 0) (KIPS)	ALLOWABLE MOMENT $M^{(1),(2)}$ (P,V,T = 0) (KIP-FT)	ALLOWABLE SHEAR $V^{(1),(2)}$ (P,M,T = 0) (KIPS)	ALLOWABLE TORSION $T^{(1),(2)}$ (P,M,V = 0) (KIP-FT)	SLENDERNESS RATIO $^{(3),(4)}$ KLu/r (KIP-FT)
32	1609.36	650.93	26.25	31.20	35.55
28	1859.37	650.93	29.37	31.20	31.11
24	1859.37	650.93	33.12	31.20	26.66
20	1859.37	650.93	38.11	31.20	22.22
16	1859.37	650.93	45.61	31.20	17.77
12	1859.37	650.93	58.12	31.20	13.33
10	1859.37	650.93	68.74	31.20	11.11

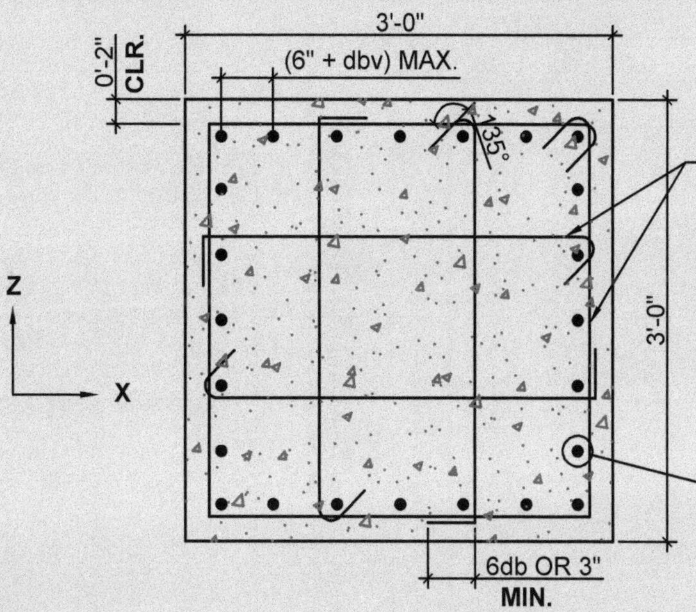

$I_{service}$ = 140107.96 IN^4
E= 3865 KSI

db = TIES DIA.
dbv = VERT. REINF. DIA.

As / Ag = 0.011

NOTE (1) : ALLOWABLE LOADS, MOMENT, SHEAR AND TORSION, ARE TAKEN AS LRFD RESULTS (PHI*Pn, PHI*Mn, PHI*Vn, OR PHI*Tn) / (Ω = 1.6) IN OTHER WORDS TO CONVERT FROM ASD TO LRFD RESULTS MULTIPLY TABULATED RESULTS ABOVE BY A FACTOR OF 1.6

NOTE (2): REINFORCING STEEL IS GR. 60 ASTM A615

NOTE (3): RADIUS OF GYRATION = 0.30*H = 10.8 INCHES, WHERE H = OVERALL DIMENSION OF A RECTANGULAR COLUMN IN THE DIRECTION OF THE MOMENT

NOTE (4): TABULATED RESULTS ARE BASED ON K = EFFECTIVE LENGTH FACTOR FOR COMPRESSION MEMBERS = 1.0

TABLE 4-5

NOTE: THIS CONC. PIER HAVE CLOSER TIES SPACING TO MAXIMIZE ALLOWABLE SHEAR

36"x36" REINFORCED CONCRETE PIERS					
f'c = 4500 PSI , NORMAL WEIGHT CONC. (WT = 0.145 K/FT^3)					
HEIGHT (FEET)	ALLOWABLE AXIAL LOADS P (1),(2) (M,V,T = 0) (KIPS)	ALLOWABLE MOMENT M (1),(2) (P,V,T = 0) (KIP-FT)	ALLOWABLE SHEAR V (1),(2) (P,M,T = 0) (KIPS)	ALLOWABLE TORSION T (1),(2) (P,M,V = 0) (KIP-FT)	SLENDERNESS RATIO (3),(4) KLu/r (KIP-FT)
32	1609.36	650.93	32.14	31.20	35.55
28	1859.37	650.93	35.97	31.20	31.11
24	1859.37	650.93	40.56	31.20	26.66
20	1859.37	650.93	46.67	31.20	22.22
16	1859.37	650.93	55.86	31.20	17.77
12	1859.37	650.93	71.18	31.20	13.33
10	1859.37	650.93	84.18	31.20	11.11

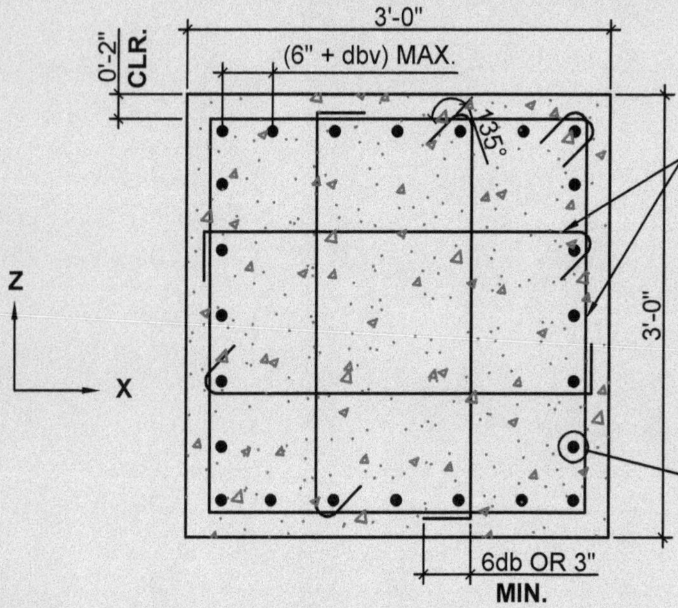

$I_{service}$ = 140107.96 IN^4
E= 3865 KSI

db = TIES DIA.
dbv = VERT. REINF. DIA.

As / Ag = 0.011

NOTE (1) : ALLOWABLE LOADS, MOMENT, SHEAR AND TORSION, ARE TAKEN AS LRFD RESULTS (PHI*Pn, PHI*Mn, PHI*Vn, OR PHI*Tn) / (Ω = 1.6) IN OTHER WORDS TO CONVERT FROM ASD TO LRFD RESULTS MULTIPLY TABULATED RESULTS ABOVE BY A FACTOR OF 1.6

NOTE (2): REINFORCING STEEL IS GR. 60 ASTM A615

NOTE (3): RADIUS OF GYRATION = 0.30*H = 10.8 INCHES, WHERE H = OVERALL DIMENSION OF A RECTANGULAR COLUMN IN THE DIRECTION OF THE MOMENT

NOTE (4): TABULATED RESULTS ARE BASED ON K = EFFECTIVE LENGTH FACTOR FOR COMPRESSION MEMBERS = 1.0

TABLE 4-6

36"x24" REINFORCED CONCRETE PIERS					
f'c = 3500 PSI , NORMAL WEIGHT CONC. (WT = 0.145 K/FT^3)					
HEIGHT (FEET)	ALLOWABLE AXIAL LOADS P$^{(1),(2)}$ (M,V,T = 0) (KIPS)	ALLOWABLE MOMENT M $^{(1),(2),(5)}$ (P,V,T = 0) (KIP-FT)	ALLOWABLE SHEAR V $^{(1),(2),(5)}$ (P,M,T = 0) (KIPS)	ALLOWABLE TORSION T $^{(1),(2)}$ (P,M,V = 0) (KIP-FT)	SLENDERNESS RATIO $^{(3),(4)}$ KLu/r (KIP-FT)
32	390.62	314.12	15.62	14.37	53.33
28	786.87	314.12	16.87	14.37	46.66
24	950.00	314.12	18.12	14.37	40.00
20	1028.12	314.12	20.31	14.37	33.33
16	1028.12	314.12	23.43	14.37	26.66
12	1028.12	314.12	29.37	14.37	20.00
10	1028.12	314.12	33.75	14.37	16.66

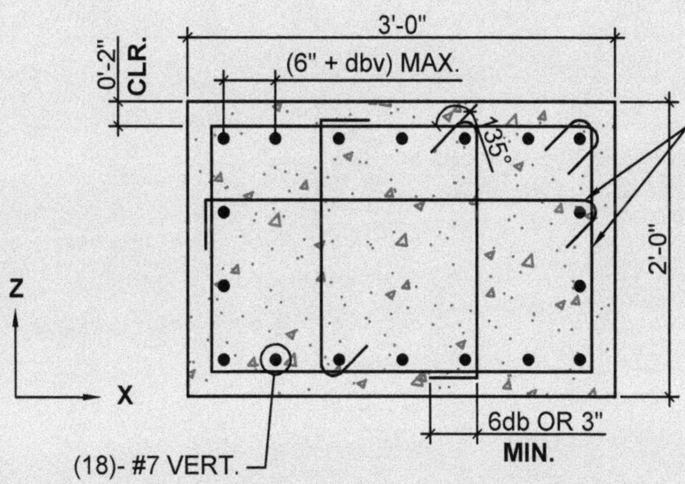

$I_{service}$ = 93405.31 IN^4
E= 3409 KSI

db = TIES DIA.
dbv = VERT. REINF. DIA.

As / Ag = 0.0125

NOTE (1) : ALLOWABLE LOADS, MOMENT, SHEAR AND TORSION, ARE TAKEN AS LRFD RESULTS (PHI*Pn, PHI*Mn, PHI*Vn, OR PHI*Tn) / (Ω = 1.6) IN OTHER WORDS TO CONVERT FROM ASD TO LRFD RESULTS MULTIPLY TABULATED RESULTS ABOVE BY A FACTOR OF 1.6

NOTE (2): REINFORCING STEEL IS GR. 60 ASTM A615

NOTE (3): RADIUS OF GYRATION = 0.30*H = 7.2 INCHES, WHERE H = OVERALL DIMENSION OF A RECTANGULAR COLUMN IN THE DIRECTION OF THE MOMENT

NOTE (4): TABULATED RESULTS ARE BASED ON K = EFFECTIVE LENGTH FACTOR FOR COMPRESSION MEMBERS = 1.0

TABLE 4-7

36"x24" REINFORCED CONCRETE PIERS					
f'c = 4000 PSI , NORMAL WEIGHT CONC. (WT = 0.145 K/FT^3)					
HEIGHT (FEET)	ALLOWABLE AXIAL LOADS P (1),(2) (M,V,T = 0) (KIPS)	ALLOWABLE MOMENT M (1),(2),(5) (P,V,T = 0) (KIP-FT)	ALLOWABLE SHEAR V (1),(2),(5) (P,M,T = 0) (KIPS)	ALLOWABLE TORSION T (1),(2) (P,M,V = 0) (KIP-FT)	SLENDERNESS RATIO (3),(4) KLu/r (KIP-FT)
32	515.62	315.91	15.62	15.36	53.33
28	890.62	315.91	16.87	15.36	46.66
24	1112.5	315.91	18.12	15.36	40.00
20	1140.62	315.91	20.31	15.36	33.33
16	1140.62	315.91	23.43	15.36	26.66
12	1140.62	315.91	29.37	15.36	20.00
10	1140.62	315.91	33.75	15.36	16.66

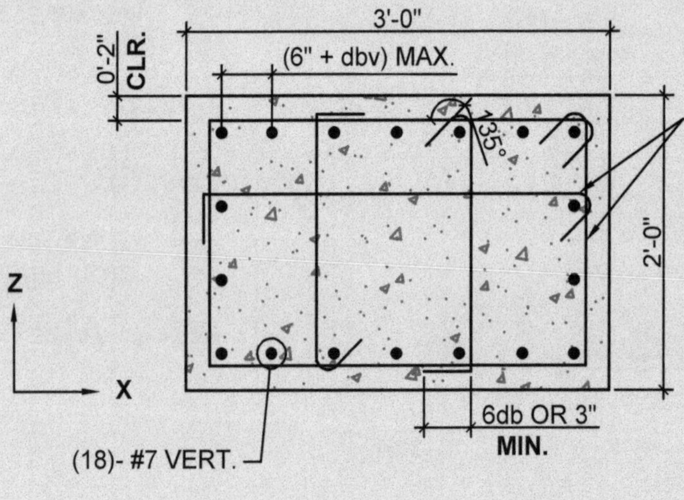

$I_{service}$ = 93405.31 IN^4
E= 3644 KSI

db = TIES DIA.
dbv = VERT. REINF. DIA.

As / Ag = 0.0125

NOTE (1) : ALLOWABLE LOADS, MOMENT, SHEAR AND TORSION, ARE TAKEN AS LRFD RESULTS (PHI*Pn, PHI*Mn, PHI*Vn, OR PHI*Tn) / (Ω = 1.6) IN OTHER WORDS TO CONVERT FROM ASD TO LRFD RESULTS MULTIPLY TABULATED RESULTS ABOVE BY A FACTOR OF 1.6

NOTE (2): REINFORCING STEEL IS GR. 60 ASTM A615

NOTE (3): RADIUS OF GYRATION = 0.30*H = 7.2 INCHES, WHERE H = OVERALL DIMENSION OF A RECTANGULAR COLUMN IN THE DIRECTION OF THE MOMENT

NOTE (4): TABULATED RESULTS ARE BASED ON K = EFFECTIVE LENGTH FACTOR FOR COMPRESSION MEMBERS = 1.0

TABLE 4-8

36"x24" REINFORCED CONCRETE PIERS					
f'c = 4500 PSI , NORMAL WEIGHT CONC. (WT = 0.145 K/FT^3)					
HEIGHT (FEET)	ALLOWABLE AXIAL LOADS P (1),(2) (M,V,T = 0) (KIPS)	ALLOWABLE MOMENT M (1),(2),(5) (P,V,T = 0) (KIP-FT)	ALLOWABLE SHEAR V (1),(2),(5) (P,M,T = 0) (KIPS)	ALLOWABLE TORSION T (1),(2) (P,M,V = 0) (KIP-FT)	SLENDERNESS RATIO (3),(4) KLu/r (KIP-FT)
32	640.62	317.65	15.62	17.38	53.33
28	1034.37	317.65	16.87	17.38	46.66
24	1250.00	317.65	18.12	17.38	40.00
20	1262.5	317.65	20.31	17.38	33.33
16	1262.5	317.65	23.43	17.38	26.66
12	1262.5	317.65	29.37	17.38	20.00
10	1262.5	317.65	33.75	17.38	16.66

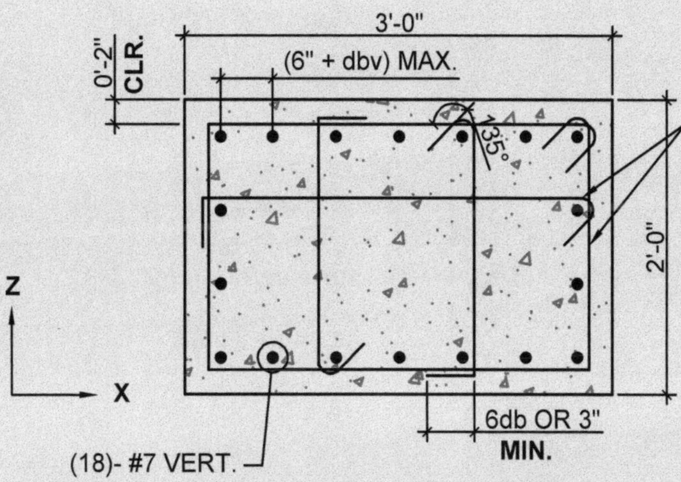

$I_{service}$ = 93405.31 IN^4
E= 3865 KSI

db = TIES DIA.
dbv = VERT. REINF. DIA.

As / Ag = 0.0125

NOTE (1) : ALLOWABLE LOADS, MOMENT, SHEAR AND TORSION, ARE TAKEN AS LRFD RESULTS (PHI*Pn, PHI*Mn, PHI*Vn, OR PHI*Tn) / (Ω = 1.6) IN OTHER WORDS TO CONVERT FROM ASD TO LRFD RESULTS MULTIPLY TABULATED RESULTS ABOVE BY A FACTOR OF 1.6

NOTE (2): REINFORCING STEEL IS GR. 60 ASTM A615

NOTE (3): RADIUS OF GYRATION = 0.30*H = 7.2 INCHES, WHERE H = OVERALL DIMENSION OF A RECTANGULAR COLUMN IN THE DIRECTION OF THE MOMENT

NOTE (4): TABULATED RESULTS ARE BASED ON K = EFFECTIVE LENGTH FACTOR FOR COMPRESSION MEMBERS = 1.0

TABLE 4-9

24"x24" REINFORCED CONCRETE PIERS					
f'c = 3500 PSI , NORMAL WEIGHT CONC. (WT = 0.145 K/FT^3)					
HEIGHT (FEET)	ALLOWABLE AXIAL LOADS P(1),(2) (M,V,T = 0) (KIPS)	ALLOWABLE MOMENT M (1),(2),(5) (P,V,T = 0) (KIP-FT)	ALLOWABLE SHEAR V (1),(2),(5) (P,M,T = 0) (KIPS)	ALLOWABLE TORSION T (1),(2) (P,M,V = 0) (KIP-FT)	SLENDERNESS RATIO (3),(4) KLu/r (KIP-FT)
32	302.64	215.25	6.29	7.98	53.33
28	415.93	215.25	7.63	7.98	46.66
24	625.42	215.25	9.50	7.98	40.00
20	652.05	215.25	12.24	7.98	33.33
16	652.05	215.25	14.40	7.98	26.66
12	652.05	215.25	18.15	7.98	20.00
10	652.05	215.25	21.52	7.98	16.66

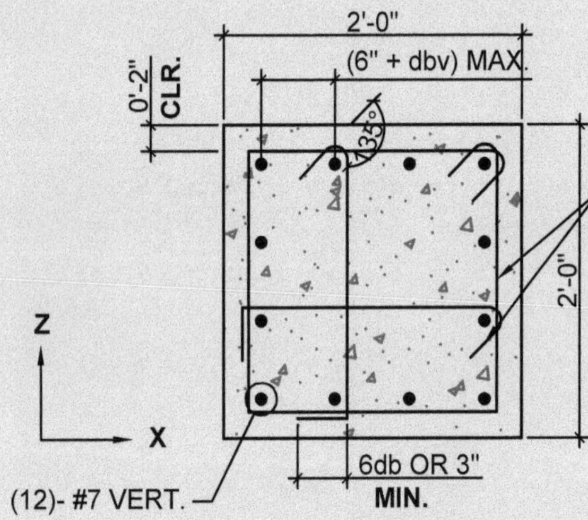

$I_{service}$ = 27675.64 IN^4
E= 3409 KSI

db = TIES DIA.
dbv = VERT. REINF. DIA.

As / Ag = 0.0125

NOTE (1) : ALLOWABLE LOADS, MOMENT, SHEAR AND TORSION, ARE TAKEN AS LRFD RESULTS (PHI*Pn, PHI*Mn, PHI*Vn, OR PHI*Tn) / (Ω = 1.6) IN OTHER WORDS TO CONVERT FROM ASD TO LRFD RESULTS MULTIPLY TABULATED RESULTS ABOVE BY A FACTOR OF 1.6

NOTE (2): REINFORCING STEEL IS GR. 60 ASTM A615

NOTE (3): RADIUS OF GYRATION = 0.30*H = 7.2 INCHES, WHERE H = OVERALL DIMENSION OF A RECTANGULAR COLUMN IN THE DIRECTION OF THE MOMENT

NOTE (4): TABULATED RESULTS ARE BASED ON K = EFFECTIVE LENGTH FACTOR FOR COMPRESSION MEMBERS = 1.0

TABLE 4-10

24"x24" REINFORCED CONCRETE PIERS					
f'c = 4000 PSI , NORMAL WEIGHT CONC. (WT = 0.145 K/FT^3)					
HEIGHT (FEET)	ALLOWABLE AXIAL LOADS P$^{(1),(2)}$ (M,V,T = 0) (KIPS)	ALLOWABLE MOMENT M$^{(1),(2),(5)}$ (P,V,T = 0) (KIP-FT)	ALLOWABLE SHEAR V$^{(1),(2),(5)}$ (P,M,T = 0) (KIPS)	ALLOWABLE TORSION T$^{(1),(2)}$ (P,M,V = 0) (KIP-FT)	SLENDERNESS RATIO$^{(3),(4)}$ KLu/r (KIP-FT)
32	367.93	216.28	8.28	8.53	53.33
28	543.30	216.28	10.00	8.53	46.66
24	734.95	216.28	11.25	8.53	40.00
20	741.59	216.28	13.12	8.53	33.33
16	741.59	216.28	15.62	8.53	26.66
12	741.59	216.28	19.68	8.53	20.00
10	741.59	216.28	23.12	8.53	16.66

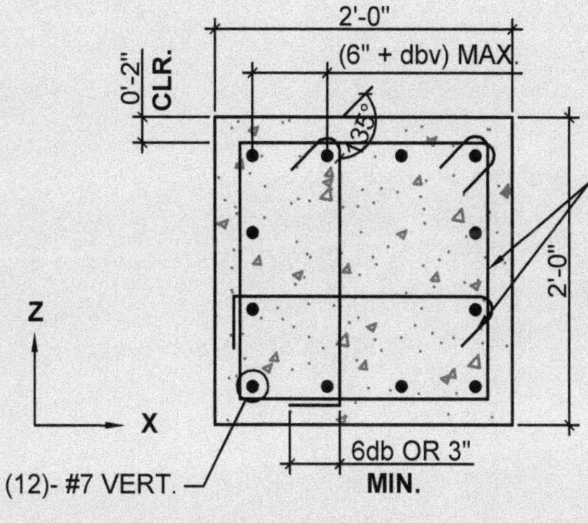

$I_{service}$ = 27675.64 IN^4
E= 3644 KSI

db = TIES DIA.
dbv = VERT. REINF. DIA.

As / Ag = 0.0125

NOTE (1) : ALLOWABLE LOADS, MOMENT, SHEAR AND TORSION, ARE TAKEN AS LRFD RESULTS (PHI*Pn, PHI*Mn, PHI*Vn, OR PHI*Tn) / (Ω = 1.6) IN OTHER WORDS TO CONVERT FROM ASD TO LRFD RESULTS MULTIPLY TABULATED RESULTS ABOVE BY A FACTOR OF 1.6

NOTE (2): REINFORCING STEEL IS GR. 60 ASTM A615

NOTE (3): RADIUS OF GYRATION = 0.30*H = 7.2 INCHES, WHERE H = OVERALL DIMENSION OF A RECTANGULAR COLUMN IN THE DIRECTION OF THE MOMENT

NOTE (4): TABULATED RESULTS ARE BASED ON K = EFFECTIVE LENGTH FACTOR FOR COMPRESSION MEMBERS = 1.0

TABLE 4-11

24"x24" REINFORCED CONCRETE PIERS					
f'c = 4500 PSI , NORMAL WEIGHT CONC. (WT = 0.145 K/FT^3)					
HEIGHT (FEET)	ALLOWABLE AXIAL LOADS P(1),(2) (M,V,T = 0) (KIPS)	ALLOWABLE MOMENT M (1),(2),(5) (P,V,T = 0) (KIP-FT)	ALLOWABLE SHEAR V (1),(2),(5) (P,M,T = 0) (KIPS)	ALLOWABLE TORSION T (1),(2) (P,M,V = 0) (KIP-FT)	SLENDERNESS RATIO (3),(4) KLu/r (KIP-FT)
32	383.43	217.30	8.28	9.05	53.33
28	639.49	217.30	10.00	9.05	46.66
24	806.79	217.30	11.25	9.05	40.00
20	813.09	217.30	13.12	9.05	33.33
16	813.09	217.30	15.62	9.05	26.66
12	813.09	217.30	19.68	9.05	20.00
10	813.09	217.30	23.12	9.05	16.66

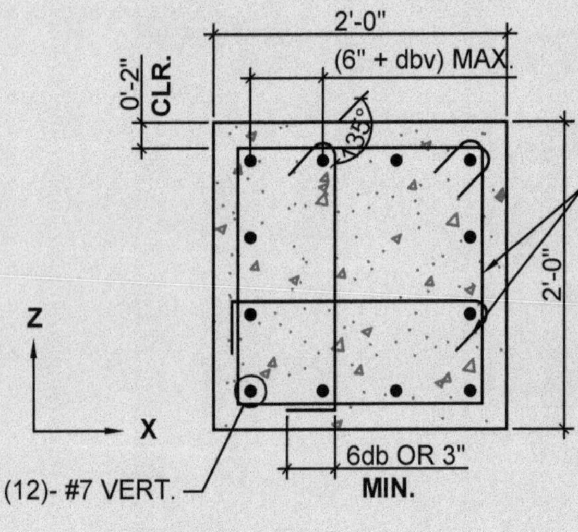

$I_{service}$ = 27675.64 IN^4
E= 3865 KSI

db = TIES DIA.
dbv = VERT. REINF. DIA.

As / Ag = 0.0125

NOTE (1) : ALLOWABLE LOADS, MOMENT, SHEAR AND TORSION, ARE TAKEN AS LRFD RESULTS (PHI*Pn, PHI*Mn, PHI*Vn, OR PHI*Tn) / (Ω = 1.6) IN OTHER WORDS TO CONVERT FROM ASD TO LRFD RESULTS MULTIPLY TABULATED RESULTS ABOVE BY A FACTOR OF 1.6

NOTE (2): REINFORCING STEEL IS GR. 60 ASTM A615

NOTE (3): RADIUS OF GYRATION = 0.30*H = 7.2 INCHES, WHERE H = OVERALL DIMENSION OF A RECTANGULAR COLUMN IN THE DIRECTION OF THE MOMENT

NOTE (4): TABULATED RESULTS ARE BASED ON K = EFFECTIVE LENGTH FACTOR FOR COMPRESSION MEMBERS = 1.0

TABLE 4-12

14"x14" CONCRETE CAPACITY AT VARIOUS HEIGHTS [1,2,3]			
CONCRETE COMPRESSIVE STRENGTH (f'c) = 3500 PSI, NORMAL WEIGHT CONCRETE (Λ = 1), DENSITY = 145 PC			
APPLIED UNFACTORED VERTICAL LOAD (P_{AXIAL}) (KIPS)	APPLIED UNFACTORED MOMENT LOAD IN ANY DIRECTION (M_{Z-Z}, M_{X-X}, M_{Y-Y}) (KIP-FT)	APPLIED UNFACTORED SHEAR LOAD IN ANY DIRECTION (V_{Z-Z}, V_{X-X}) (KIP-FT)	MAXIMUM ALLOWABLE HEIGHT BASED ON LOADING CRITERIA (INCHES)
276	0	0	280
0	42.4	0	118
0	0	6.7	118
137	21.3	6.7	84
148	26.4	7.1	72
158	29.3	7.8	60
166	32.6	8.7	48
177	36.3	9.7	36
182	40.4	11.2	24

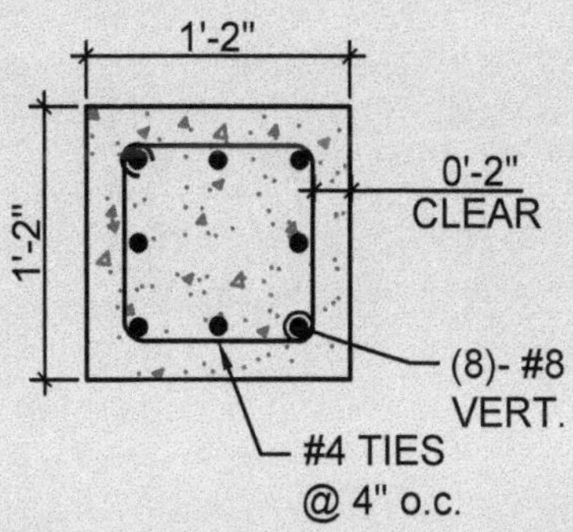

NOTE (1): PIER MINIMUM COVER = 2 INCHES.

NOTE (2): SEE ABOVE FOR REINFORCEMENT DETAIL AND TIES SPACING, THIS APPLIES TO ALL HEIGHTS

NOTE (3) : ALL TABULATED RESULTS ARE BASED ON APPLIED LIVE LOAD (WORST CASE SCENARIO FOR LRFD LOAD COMBINATION)

NOTE (4): PIER OWN WEIGHT HAS ALREADY BEEN FACTORED INTO THE CALCULATIONS.

NOTE (6): STEEL REINFORCEMENT IS GRADE 60 STEEL (ASTM A615)

TABLE 4-13

12"x12" CONCRETE CAPACITY AT VARIOUS HEIGHTS [1,2,3]			
CONCRETE COMPRESSIVE STRENGTH (f'c) = 3500 PSI, NORMAL WEIGHT CONCRETE (λ = 1), DENSITY = 145 PC			
APPLIED UNFACTORED VERTICAL LOAD (P_{AXIAL}) (KIPS)	APPLIED UNFACTORED MOMENT LOAD IN ANY DIRECTION (M_{Z-Z}, M_{X-X}, M_{Y-Y}) (KIP-FT)	APPLIED UNFACTORED SHEAR LOAD IN ANY DIRECTION (V_{Z-Z}, V_{X-X}) (KIP-FT)	MAXIMUM ALLOWABLE HEIGHT BASED ON LOADING CRITERIA (INCHES)
240	0	0	240
0	41.6	0	98
0	0	5.1	98
117	3.5	5.1	80
132	9.5	5.1	72
132	14.0	5.6	60
142	18.0	5.9	48
155	22	6.1	36
160	25.0	7.1	24

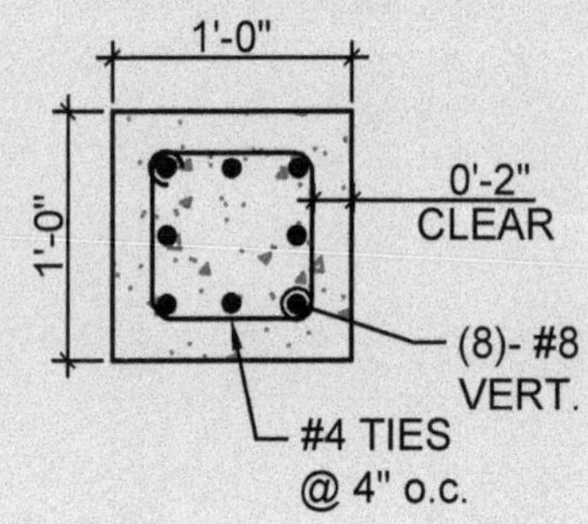

NOTE (1): PIER MINIMUM COVER = 2 INCHES.

NOTE (2): SEE ABOVE FOR REINFORCEMENT DETAIL AND TIES SPACING, THIS APPLIES TO ALL HEIGHTS

NOTE (3) : ALL TABULATED RESULTS ARE BASED ON APPLIED LIVE LOAD (WORST CASE SCENARIO FOR LRFD LOAD COMBINATION)

NOTE (4): PIER OWN WEIGHT HAS ALREADY BEEN FACTORED INTO THE CALCULATIONS.

NOTE (6): STEEL REINFORCEMENT IS GRADE 60 STEEL (ASTM A615)

MASONRY WALLS AND MASONRY BOND BEAMS CAPACITY

TABLE 5-1

ALLOWABLE CONCRETE MASONRY WALLS CAPACITY							
8-INCH CMU WALL (1), (2), (3) (MAXIMUM HEIGHT = 13'-4")							
REBAR SIZE AND SPACING (IN GROUTED CORE)	A_N IN^2/FT	A_{ST} IN^2/FT	I IN^4/FT	S IN^3/FT	ALLOWABLE COMPRESSIVE FORCE (KIPS/FT)	ALLOWABLE SHEAR FORCE (KIPS/FT)	ALLOWABLE MOMENT (KIP-FT/FT)
#5 @ 96" o.c	37.9	0.038	337.1	88.4	15.50	0.55	0.36
#5 @ 72" o.c	39.8	0.051	340.8	89.4	16.20	0.73	0.41
#5 @ 48" o.c	42.8	0.077	346.8	91.0	17.23	1.09	0.62
#5 @ 40" o.c	44.7	0.092	350.6	92.0	17.88	1.15	0.74
#5 @ 32" o.c	47.6	0.115	356.4	93.5	18.86	1.26	0.92
#5 @ 24" o.c	52.5	0.153	366.1	96.0	20.48	1.38	1.18
#5 @ 16" o.c	62.3	0.230	385.4	101.1	23.59	1.66	1.35
#5 @ 8" o.c	91.5	0.460	443.3	116.3	32.08	2.45	1.67
12-INCH CMU WALL (1), (2), (3) (MAXIMUM HEIGHT = 20'-0")							
REBAR SIZE AND SPACING (IN GROUTED CORE)	A_N IN^2/FT	A_{ST} IN^2/FT	I IN^4/FT	S IN^3/FT	ALLOWABLE COMPRESSIVE FORCE (KIPS/FT)	ALLOWABLE SHEAR FORCE (KIPS/FT)	ALLOWABLE MOMENT (KIP-FT/FT)
#5 @ 96" o.c	47.1	0.038	1434.5	210.6	20.41	0.86	0.48
#5 @ 72" o.c	50.8	0.051	1469.4	215.7	21.91	1.14	0.64
#5 @ 48" o.c	58.2	0.077	1539.2	225.9	24.83	1.31	0.95
#5 @ 40" o.c	62.7	0.092	1581.0	232.1	26.59	1.41	1.14
#5 @ 32" o.c	69.4	0.115	1643.8	241.3	29.2	1.60	1.50
#5 @ 24" o.c	80.5	0.153	1748.5	256.7	33.41	1.82	1.87
#5 @ 16" o.c	102.7	0.230	1957.7	287.4	41.70	2.77	2.65
#5 @ 8" o.c	163.5	0.460	2529.4	371.3	63.57	3.74	3.44

NOTE (1) : F_M = 1500 PSI
NOTE (2) : ASSUME GRADE 60 STEEL IS USED
NOTE (3) : FACTORS OF SAFETY USED :

F_b = (1/3) F_M = 500 PSI
F_S = 24,000 PSI
F_v = $(F_M)^{1/2}$

TABLE 5-2

ALLOWABLE SHEAR AND MOMENT CAPACITY FOR CONCRETE MASONRY LINTELS			
8-INCH CMU (1),(2),(3)			
LINTEL SIZE	STEEL REINFORCEMENT	ALLOWABLE SHEAR (KIPS)	ALLOWABLE MOMENT (KIP-FT)
NOMINAL 8x8	(1) - #4	1.59	1.47
NOMINAL 8x8	(1) - #5	1.57	1.66
NOMINAL 8x8	(2) - #4	1.59	1.82
NOMINAL 8x8	(2) - #5	1.57	2.00
NOMINAL 8x16	(1) - #5	3.93	7.42
NOMINAL 8x16	(2) - #5	3.93	9.71
NOMINAL 8x24	(1) - #5	6.29	12.12
NOMINAL 8x24	(2) - #5	6.29	21.25
12-INCH CMU (1),(2),(3)			
LINTEL SIZE	STEEL REINFORCEMENT	ALLOWABLE SHEAR (KIPS)	ALLOWABLE MOMENT (KIP-FT)
NOMINAL 12x8	(1) - #4	2.42	1.93
NOMINAL 12x8	(1) - #5	2.39	2.21
NOMINAL 12x8	(2) - #4	2.42	2.45
NOMINAL 12x8	(2) - #5	2.39	2.74
NOMINAL 12x16	(1) - #5	5.99	7.56
NOMINAL 12x16	(2) - #5	5.99	12.86
NOMINAL 12x24	(1) - #5	9.60	12.30
NOMINAL 12x24	(2) - #5	9.60	23.97

NOTE (1) : F_M = 1500 PSI
NOTE (2) : ASSUME GRADE 60 STEEL IS USED
NOTE (3) : ASSUME BOTTOM COVER = 2 INCHES

TABLE 5-3

CONCRETE MASONRY UNIT WALL WEIGHT (SINGLE WHYTHE WALLS)			
8-INCH CMU [1]		12- INCH CMU [1]	
VERTICAL GROUT SPACING (INCHES)	WALL WEIGHT (PSF)	VERTICAL GROUT SPACING (INCHES)	WALL WEIGHT (PSF)
NO GROUT	39	NO GROUT	54
96	43	96	60
72	44	72	62
48	47	48	67
40	48	40	69
32	51	32	73
24	55	24	80
16	63	16	93
8	86	8	132
NOTE (1) : WALL WEIGHS ARE BASED ON CONCRETE UNIT DENSITY OF = 135 PCF			

LIGHT GAUGE STEEL BOXED BEAMS CAPACITY

TABLE 6-1

BOXED LGS (LIGHT GAUGE STEEL) BEAM CAPACITY				

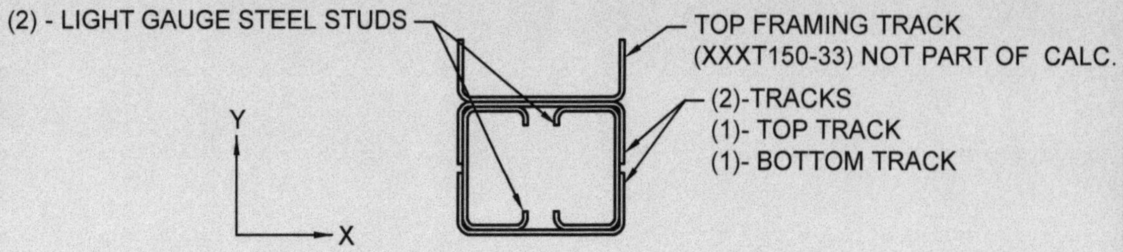

SIZE OF MEMBERS	Ixx (IN^4)	Iyy (IN^4)	ALLOWABLE MOMENT (Mz) (KIP-IN)	ALLOWABLE SHEAR [(1),(4)] (LBS)
3-1/2" WIDE BUILT-UP BOXED BEAM				
STUDS: (2)-250S162-33 [(2)] TRACKS: (2)-350T125-33 [(2)]	0.89	1.37	9.08	256
STUDS: (2)-250S162-43 [(2)] TRACKS: (2)-350T125-43 [(2)]	1.16	1.75	11.84	455
STUDS: (2)-250S162-54 [(3)] TRACKS: (2)-350T125-54 [(3)]	1.44	2.15	22.08	1106
STUDS: (2)-250S162-68 [(3)] TRACKS: (2)-350T125-54 [(3)]	1.60	2.40	24.54	1778
STUDS: (2)-250S162-97 [(3)] TRACKS: (2)-350T125-54 [(3)]	1.88	2.86	28.83	3671
STUDS: (2)-350S162-33 [(2)] TRACKS: (2)-350T150-33 [(2)]	1.90	1.66	14.06	232
STUDS: (2)-350S162-43 [(2)] TRACKS: (2)-350T150-43 [(2)]	2.46	2.12	18.11	419
STUDS: (2)-350S162-54 [(3)] TRACKS: (2)-350T150-54 [(3)]	3.07	2.61	34.03	1031
STUDS: (2)-350S162-68 [(3)] TRACKS: (2)-350T150-54 [(3)]	3.43	2.93	38.03	1673
STUDS: (2)-350S162-97 [(3)] TRACKS: (2)-350T150-54 [(3)]	4.06	3.54	45.01	3493

NOTE (1) : WEB CRIPPLING CONTROLS ALLOWABLE SHEAR RESULTS
NOTE (2) : F_y = 33 KSI
NOTE (3) : F_y = 50 KSI
NOTE (4): IF REQUIRED SHEAR EXCEEDS ALLOWABLE SHEAR RESULTS, WEB STIFFENERS WILL BE REQUIRED

TABLE 6-2

BOXED LGS (LIGHT GAUGE STEEL) BEAM CAPACITY

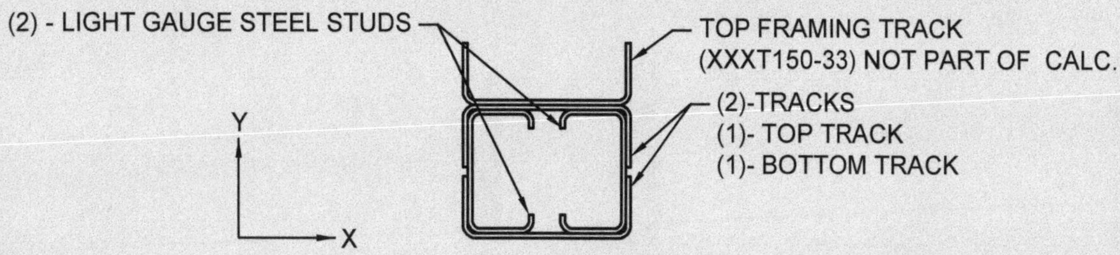

SIZE OF MEMBERS	Ixx (IN^4)	Iyy (IN^4)	ALLOWABLE MOMENT (Mz) (KIP-IN)	ALLOWABLE SHEAR [(1),(4)] (LBS)
3-1/2" WIDE BUILT-UP BOXED BEAM				
STUDS: (2)-362S162-33 [(2)] TRACKS: (2)-350T150-33 [(2)]	2.06	1.68	14.73	229
STUDS: (2)-362S162-43 [(2)] TRACKS: (2)-350T150-43 [(2)]	2.67	2.15	19.00	415
STUDS: (2)-362S162-54 [(3)] TRACKS: (2)-350T150-54 [(3)]	3.32	2.65	35.57	1022
STUDS: (2)-362S162-68 [(3)] TRACKS: (2)-350T150-54 [(3)]	3.71	2.97	39.75	1661
STUDS: (2)-362S162-97 [(3)] TRACKS: (2)-350T150-54 [(3)]	4.41	3.61	47.25	3473
STUDS: (2)-400S162-33 [(2)] TRACKS: (2)-350T150-33 [(2)]	2.57	1.75	16.68	221
STUDS: (2)-400S162-43 [(2)] TRACKS: (2)-350T150-43 [(2)]	3.34	2.24	21.58	403
STUDS: (2)-400S162-54 [(3)] TRACKS: (2)-350T150-54 [(3)]	4.16	2.76	40.51	998
STUDS: (2)-400S162-68 [(3)] TRACKS: (2)-350T150-54 [(3)]	4.65	3.12	45.28	1627
STUDS: (2)-400S162-97 [(3)] TRACKS: (2)-350T150-54 [(3)]	5.53	3.81	53.85	3415

NOTE (1) : WEB CRIPPLING CONTROLS ALLOWABLE SHEAR RESULTS
NOTE (2) : F_y = 33 KSI
NOTE (3) : F_y = 50 KSI
NOTE (4): IF REQUIRED SHEAR EXCEEDS ALLOWABLE SHEAR RESULTS, WEB STIFFENERS WILL BE REQUIRED

TABLE 6-3

BOXED LGS (LIGHT GAUGE STEEL) BEAM CAPACITY

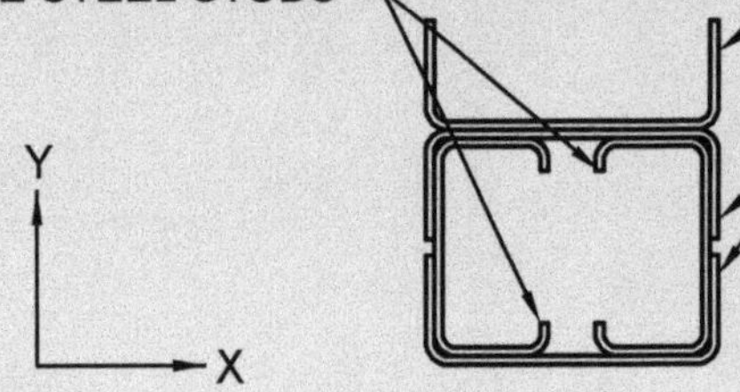

SIZE OF MEMBERS	Ixx (IN^4)	Iyy (IN^4)	ALLOWABLE MOMENT (Mz) (KIP-IN)	ALLOWABLE SHEAR [(1),(4)] (LBS)
3-1/2" WIDE BUILT-UP BOXED BEAM				
STUDS: (2)-550S162-33 [(2)] TRACKS: (2)-350T150-33 [(2)]	5.32	2.04	25.23	192
STUDS: (2)-550S162-43 [(2)] TRACKS: (2)-350T150-43 [(2)]	6.89	2.61	32.56	361
STUDS: (2)-550S162-54 [(3)] TRACKS: (2)-350T150-54 [(3)]	8.58	3.21	61.20	909
STUDS: (2)-550S162-68 [(3)] TRACKS: (2)-350T150-54 [(3)]	9.65	3.68	68.83	1504
STUDS: (2)-550S162-97 [(3)] TRACKS: (2)-350T150-54 [(3)]	11.58	4.59	82.60	3211
STUDS: (2)-600S162-33 [(2)] TRACKS: (2)-350T150-33 [(2)]	6.48	2.13	28.20	183
STUDS: (2)-600S162-43 [(2)] TRACKS: (2)-350T150-43 [(2)]	8.40	2.73	36.44	348
STUDS: (2)-600S162-54 [(3)] TRACKS: (2)-350T150-54 [(3)]	10.47	3.36	68.57	883
STUDS: (2)-600S162-68 [(3)] TRACKS: (2)-350T150-54 [(3)]	11.78	3.87	77.14	1467
STUDS: (2)-600S162-97 [(3)] TRACKS: (2)-350T150-54 [(3)]	14.19	4.86	92.93	3150

NOTE (1) : WEB CRIPPLING CONTROLS ALLOWABLE SHEAR RESULTS
NOTE (2) : F_y = 33 KSI
NOTE (3) : F_y = 50 KSI
NOTE (4): IF REQUIRED SHEAR EXCEEDS ALLOWABLE SHEAR RESULTS, WEB STIFFENERS WILL BE REQUIRED

TABLE 6-4

BOXED LGS (LIGHT GAUGE STEEL) BEAM CAPACITY				
SIZE OF MEMBERS	Ixx (IN^4)	Iyy (IN^4)	ALLOWABLE MOMENT (Mz) (KIP-IN)	ALLOWABLE SHEAR [(1),(4)] (LBS)
3-1/2" WIDE BUILT-UP BOXED BEAM				
STUDS: (2)-800S162-43 [(2)] TRACKS: (2)-350T150-43 [(2)]	16.26	3.22	53.08	302
STUDS: (2)-800S162-54 [(3)] TRACKS: (2)-350T150-54 [(3)]	20.26	3.96	99.95	787
STUDS: (2)-800S162-68 [(3)] TRACKS: (2)-350T150-54 [(3)]	22.92	4.62	113.07	1335
STUDS: (2)-800S162-97 [(3)] TRACKS: (2)-350T150-54 [(3)]	27.87	5.91	137.50	2929
STUDS: (2)-1000S162-54 [(3)] TRACKS: (2)-350T200-54 [(3)]	33.95	4.56	134.35	702
STUDS: (2)-1000S162-68 [(3)] TRACKS: (2)-350T200-54 [(3)]	38.60	5.37	152.75	1218
STUDS: (2)-1000S162-97 [(3)] TRACKS: (2)-350T200-54 [(3)]	47.37	6.95	187.47	2737
STUDS: (2)-1200S162-68 [(3)] TRACKS: (2)-350T200-54 [(3)]	59.36	6.11	196.10	1114
STUDS: (2)-1200S162-97 [(3)] TRACKS: (2)-350T200-54 [(3)]	73.44	8.00	242.62	2564
STUDS: (2)-1400S162-68 [(3)] TRACKS: (2)-350T300-54 [(3)]	85.75	6.86	243.98	1018

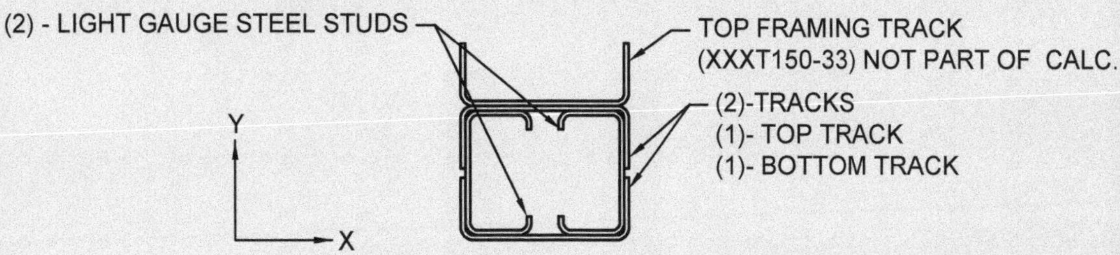

NOTE (1) : WEB CRIPPLING CONTROLS ALLOWABLE SHEAR RESULTS
NOTE (2) : F_y = 33 KSI
NOTE (3) : F_y = 50 KSI
NOTE (4): IF REQUIRED SHEAR EXCEEDS ALLOWABLE SHEAR RESULTS, WEB STIFFENERS WILL BE REQUIRED

TABLE 6-5

BOXED LGS (LIGHT GAUGE STEEL) BEAM CAPACITY

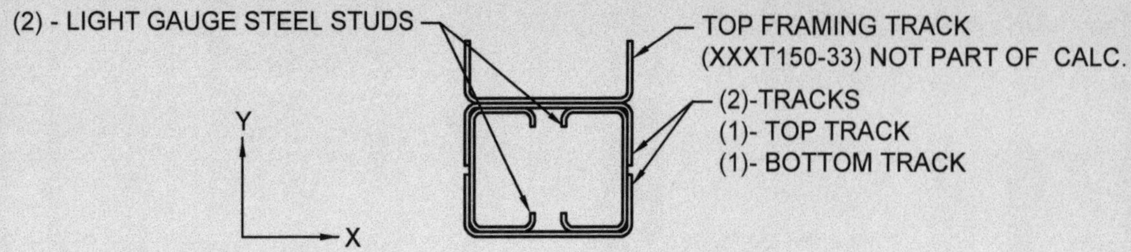

SIZE OF MEMBERS	Ixx (IN^4)	Iyy (IN^4)	ALLOWABLE MOMENT (Mz) (KIP-IN)	ALLOWABLE SHEAR [(1),(4)] (LBS)
3-1/2" WIDE BUILT-UP BOXED BEAM				
STUDS: (2)-1400S162-97 [(3)] TRACKS: (2)-350T300-54 [(3)]	106.86	9.05	302.98	2406
6" WIDE BUILT-UP BOXED BEAM				
STUDS: (2)-250S162-33 [(2)] TRACKS: (2)-600T125-33 [(2)]	1.15	5.05	9.86	256
STUDS: (2)-250S162-43 [(2)] TRACKS: (2)-600T125-43 [(2)]	1.50	6.50	12.76	455
STUDS: (2)-250S162-54 [(3)] TRACKS: (2)-600T125-54 [(3)]	1.88	8.04	24.03	1106
STUDS: (2)-250S162-68 [(3)] TRACKS: (2)-600T125-54 [(3)]	2.04	9.00	26.07	1778
STUDS: (2)-250S162-97 [(3)] TRACKS: (2)-600T125-54 [(3)]	2.32	10.80	29.65	3671
STUDS: (2)-350S162-33 [(2)] TRACKS: (2)-600T150-33 [(2)]	2.41	5.92	14.86	232
STUDS: (2)-350S162-43 [(2)] TRACKS: (2)-600T150-43 [(2)]	3.14	7.62	19.26	419
STUDS: (2)-350S162-54 [(3)] TRACKS: (2)-600T150-54 [(3)]	3.92	9.44	36.21	1031

NOTE (1) : WEB CRIPPLING CONTROLS ALLOWABLE SHEAR RESULTS
NOTE (2) : F_y = 33 KSI
NOTE (3) : F_y = 50 KSI
NOTE (4): IF REQUIRED SHEAR EXCEEDS ALLOWABLE SHEAR RESULTS, WEB STIFFENERS WILL BE REQUIRED

TABLE 6-6

BOXED LGS (LIGHT GAUGE STEEL) BEAM CAPACITY				

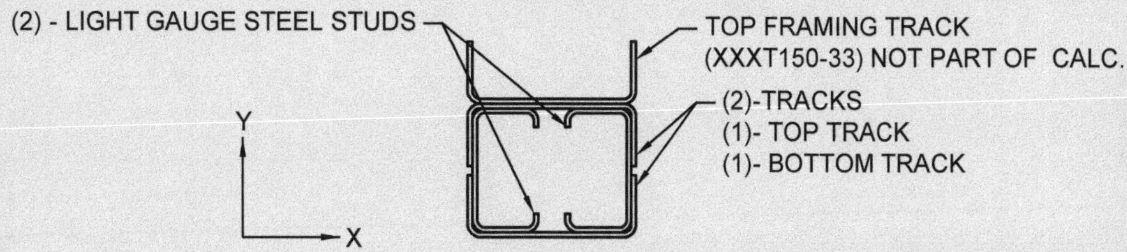

SIZE OF MEMBERS	Ixx (IN^4)	Iyy (IN^4)	ALLOWABLE MOMENT (Mz) (KIP-IN)	ALLOWABLE SHEAR [(1),(4)] (LBS)
6" WIDE BUILT-UP BOXED BEAM				
STUDS: (2)-350S162-68 [(3)] TRACKS: (2)-600T150-54 [(3)]	3.43	2.93	38.03	1673
STUDS: (2)-350S162-97 [(3)] TRACKS: (2)-600T150-54 [(3)]	4.06	3.54	45.01	3493
STUDS: (2)-362S162-33 [(2)] TRACKS: (2)-600T150-33 [(2)]	2.06	1.68	14.73	229
STUDS: (2)-362S162-43 [(2)] TRACKS: (2)-600T150-43 [(2)]	2.67	2.15	19.00	415
STUDS: (2)-362S162-54 [(3)] TRACKS: (2)-600T150-54 [(3)]	3.32	2.65	35.57	1022
STUDS: (2)-362S162-68 [(3)] TRACKS: (2)-600T150-54 [(3)]	3.71	2.97	39.75	1661
STUDS: (2)-362S162-97 [(3)] TRACKS: (2)-600T150-54 [(3)]	4.41	3.61	47.25	3473
STUDS: (2)-400S162-33 [(2)] TRACKS: (2)-600T150-33 [(2)]	2.57	1.75	16.68	221
STUDS: (2)-400S162-43 [(2)] TRACKS: (2)-600T150-43 [(2)]	3.34	2.24	21.58	403
STUDS: (2)-400S162-54 [(3)] TRACKS: (2)-600T150-54 [(3)]	4.16	2.76	40.51	998

NOTE (1) : WEB CRIPPLING CONTROLS ALLOWABLE SHEAR RESULTS
NOTE (2) : F_y = 33 KSI
NOTE (3) : F_y = 50 KSI
NOTE (4): IF REQUIRED SHEAR EXCEEDS ALLOWABLE SHEAR RESULTS, WEB STIFFENERS WILL BE REQUIRED

SAWN CUT WOOD BEAM AND POSTS STRUCTURAL CAPACITY

TABLE 7-1

SAWN CUT LUMBER BEAMS / GIRDERS STRUCTURAL CAPACITY					
#2 ALASKA CEDAR					
SIZE OF MEMBER	ALLOWABLE MOMENT (1),(3) (KIP-FT)	SIZE OF MEMBER	ALLOWABLE MOMENT (1),(3) (KIP-FT)	SIZE OF MEMBER	ALLOWABLE MOMENT (1),(3) (KIP-FT)
2x4	0.35	(2)-2x4's	0.70	(3)-2x4's	1.05
2x6	0.74	(2)-2x6's	1.50	(3)-2x6's	2.25
2x8	1.19	(2)-2x8's	2.41	(3)-2x8's	3.62
2x10	1.78	(2)-2x10's	3.59	(3)-2x10's	5.40
2x12	2.39	(2)-2x12's	4.83	(3)-2x12's	7.26
2x14	2.98	(2)-2x14's	6.03	(3)-2x14's	9.06
#2 ALASKA HEMLOCK					
SIZE OF MEMBER	ALLOWABLE MOMENT (1),(3) (KIP-FT)	SIZE OF MEMBER	ALLOWABLE MOMENT (1),(3) (KIP-FT)	SIZE OF MEMBER	ALLOWABLE MOMENT (1),(3) (KIP-FT)
2x4	0.33	(2)-2x4's	0.68	(3)-2x4's	1.02
2x6	0.72	(2)-2x6's	1.45	(3)-2x6's	2.18
2x8	1.15	(2)-2x8's	2.33	(3)-2x8's	3.51
2x10	1.72	(2)-2x10's	3.48	(3)-2x10's	5.23
2x12	2.31	(2)-2x12's	4.68	(3)-2x12's	7.04
2x14	3.09	(2)-2x14's	6.23	(3)-2x14's	9.37

NOTE (1) : FACTORS USED CD = 1.0, Cr = 1.15
NOTE (2): N.A. INDICATES THAT SIZE OF LUMBER IS NOT COMMONLY AVAILABLE

NOTE (3): MEMBER OWN WEIGHT HAS BEEN FACTORED INTO THE CALCULATIONS

NOTE (4): #2 WOOD IS THE MOST COMMONLY AVAILABLE GRADE COMMERCIALLY

TABLE 7-2

SAWN CUT LUMBER BEAMS / GIRDERS STRUCTURAL CAPACITY					
#2 ALASKA SPRUCE					
SIZE OF MEMBER	ALLOWABLE MOMENT (1),(3) (KIP-FT)	SIZE OF MEMBER	ALLOWABLE MOMENT (1),(3) (KIP-FT)	SIZE OF MEMBER	ALLOWABLE MOMENT (1),(3) (KIP-FT)
2x4	0.38	(2)-2x4's	0.76	(3)-2x4's	1.15
2x6	0.81	(2)-2x6's	1.64	(3)-2x6's	2.47
2x8	1.31	(2)-2x8's	2.63	(3)-2x8's	3.96
2x10	1.95	(2)-2x10's	3.93	(3)-2x10's	5.91
2x12	2.62	(2)-2x12's	5.29	(3)-2x12's	7.95
2x14	3.27	(2)-2x14's	6.60	(3)-2x14's	9.93
#2 ALASKA YELLOW CEDAR					
SIZE OF MEMBER	ALLOWABLE MOMENT (1),(3) (KIP-FT)	SIZE OF MEMBER	ALLOWABLE MOMENT (1),(3) (KIP-FT)	SIZE OF MEMBER	ALLOWABLE MOMENT (1),(3) (KIP-FT)
2x4	0.35	(2)-2x4's	0.70	(3)-2x4's	1.05
2x6	0.74	(2)-2x6's	1.50	(3)-2x6's	2.26
2x8	1.19	(2)-2x8's	2.41	(3)-2x8's	3.62
2x10	1.78	(2)-2x10's	3.59	(3)-2x10's	5.40
2x12	2.39	(2)-2x12's	4.84	(3)-2x12's	7.26
2x14	N.A. (2)	(2)-2x14's	N.A. (2)	(3)-2x14's	N.A. (2)

NOTE (1) : FACTORS USED CD = 1.0, Cr = 1.15
NOTE (2): N.A. INDICATES THAT SIZE OF LUMBER IS NOT COMMONLY AVAILABLE

NOTE (3): MEMBER OWN WEIGHT HAS BEEN FACTORED INTO THE CALCULATIONS

NOTE (4): #2 WOOD IS THE MOST COMMONLY AVAILABLE GRADE COMMERCIALLY

TABLE 7-3

SAWN CUT LUMBER BEAMS / GIRDERS STRUCTURAL CAPACITY					
#2 ASPEN					
SIZE OF MEMBER	ALLOWABLE MOMENT (1),(3) (KIP-FT)	SIZE OF MEMBER	ALLOWABLE MOMENT (1),(3) (KIP-FT)	SIZE OF MEMBER	ALLOWABLE MOMENT (1),(3) (KIP-FT)
2x4	0.26	(2)-2x4's	0.52	(3)-2x4's	0.79
2x6	0.56	(2)-2x6's	1.12	(3)-2x6's	1.69
2x8	0.89	(2)-2x8's	1.80	(3)-2x8's	2.71
2x10	1.33	(2)-2x10's	2.69	(3)-2x10's	4.05
2x12	1.79	(2)-2x12's	3.63	(3)-2x12's	5.45
2x14	2.24	(2)-2x14's	4.53	(3)-2x14's	6.81
#2 BALDCYPRESS					
SIZE OF MEMBER	ALLOWABLE MOMENT (1),(3) (KIP-FT)	SIZE OF MEMBER	ALLOWABLE MOMENT (1),(3) (KIP-FT)	SIZE OF MEMBER	ALLOWABLE MOMENT (1),(3) (KIP-FT)
2x4	0.36	(2)-2x4's	0.72	(3)-2x4's	1.08
2x6	0.77	(2)-2x6's	1.55	(3)-2x6's	2.32
2x8	1.23	(2)-2x8's	2.48	(3)-2x8's	3.73
2x10	1.84	(2)-2x10's	3.71	(3)-2x10's	5.57
2x12	2.47	(2)-2x12's	4.99	(3)-2x12's	7.49
2x14	3.08	(2)-2x14's	6.23	(3)-2x14's	9.34

NOTE (1) : FACTORS USED CD = 1.0, Cr = 1.15
NOTE (2): N.A. INDICATES THAT SIZE OF LUMBER IS NOT COMMONLY AVAILABLE

NOTE (3): MEMBER OWN WEIGHT HAS BEEN FACTORED INTO THE CALCULATIONS

NOTE (4): #2 WOOD IS THE MOST COMMONLY AVAILABLE GRADE COMMERCIALLY

TABLE 7-4

SAWN CUT LUMBER BEAMS / GIRDERS STRUCTURAL CAPACITY					
#2 BEECH-BIRCH-HICKORY					
SIZE OF MEMBER	ALLOWABLE MOMENT [(1),(3)] (KIP-FT)	SIZE OF MEMBER	ALLOWABLE MOMENT [(1),(3)] (KIP-FT)	SIZE OF MEMBER	ALLOWABLE MOMENT [(1),(3)] (KIP-FT)
2x4	0.43	(2)-2x4's	0.87	(3)-2x4's	1.31
2x6	0.93	(2)-2x6's	1.88	(3)-2x6's	2.82
2x8	1.49	(2)-2x8's	3.01	(3)-2x8's	4.52
2x10	2.22	(2)-2x10's	4.49	(3)-2x10's	6.76
2x12	2.99	(2)-2x12's	6.04	(3)-2x12's	9.08
2x14	3.73	(2)-2x14's	7.53	(3)-2x14's	11.32
#2 COAST SITKA SPRUCE					
SIZE OF MEMBER	ALLOWABLE MOMENT [(1),(3)] (KIP-FT)	SIZE OF MEMBER	ALLOWABLE MOMENT [(1),(3)] (KIP-FT)	SIZE OF MEMBER	ALLOWABLE MOMENT [(1),(3)] (KIP-FT)
2x4	0.40	(2)-2x4's	0.81	(3)-2x4's	1.22
2x6	0.86	(2)-2x6's	1.74	(3)-2x6's	2.61
2x8	1.38	(2)-2x8's	2.78	(3)-2x8's	4.18
2x10	2.06	(2)-2x10's	4.15	(3)-2x10's	6.25
2x12	2.77	(2)-2x12's	5.58	(3)-2x12's	8.41
2x14	3.45	(2)-2x14's	6.98	(3)-2x14's	10.5

NOTE (1) : FACTORS USED CD = 1.0, Cr = 1.15
NOTE (2): N.A. INDICATES THAT SIZE OF LUMBER IS NOT COMMONLY AVAILABLE

NOTE (3): MEMBER OWN WEIGHT HAS BEEN FACTORED INTO THE CALCULATIONS

NOTE (4): #2 WOOD IS THE MOST COMMONLY AVAILABLE GRADE COMMERCIALLY

TABLE 7-5

SAWN CUT LUMBER BEAMS / GIRDERS STRUCTURAL CAPACITY					
#2 COTTONWOOD					
SIZE OF MEMBER	ALLOWABLE MOMENT (1),(3) (KIP-FT)	SIZE OF MEMBER	ALLOWABLE MOMENT (1),(3) (KIP-FT)	SIZE OF MEMBER	ALLOWABLE MOMENT (1),(3) (KIP-FT)
2x4	0.27	(2)-2x4's	0.54	(3)-2x4's	0.82
2x6	0.58	(2)-2x6's	1.17	(3)-2x6's	1.76
2x8	0.93	(2)-2x8's	1.88	(3)-2x8's	2.83
2x10	1.39	(2)-2x10's	2.81	(3)-2x10's	4.22
2x12	1.87	(2)-2x12's	3.78	(3)-2x12's	5.67
2x14	2.33	(2)-2x14's	4.71	(3)-2x14's	7.08

#2 DOUGLAS FIR-SOUTH					
SIZE OF MEMBER	ALLOWABLE MOMENT (1),(3) (KIP-FT)	SIZE OF MEMBER	ALLOWABLE MOMENT (1),(3) (KIP-FT)	SIZE OF MEMBER	ALLOWABLE MOMENT (1),(3) (KIP-FT)
2x4	0.37	(2)-2x4's	0.74	(3)-2x4's	1.12
2x6	0.79	(2)-2x6's	1.59	(3)-2x6's	2.4
2x8	1.27	(2)-2x8's	2.56	(3)-2x8's	3.85
2x10	1.89	(2)-2x10's	3.82	(3)-2x10's	5.74
2x12	2.54	(2)-2x12's	5.13	(3)-2x12's	7.71
2x14	3.16	(2)-2x14's	6.42	(3)-2x14's	9.64

NOTE (1) : FACTORS USED CD = 1.0, Cr = 1.15
NOTE (2): N.A. INDICATES THAT SIZE OF LUMBER IS NOT COMMONLY AVAILABLE

NOTE (3): MEMBER OWN WEIGHT HAS BEEN FACTORED INTO THE CALCULATIONS

NOTE (4): #2 WOOD IS THE MOST COMMONLY AVAILABLE GRADE COMMERCIALLY

TABLE 7-6

SAWN CUT LUMBER BEAMS / GIRDERS STRUCTURAL CAPACITY					
#2 DOUGLAS FIR-LARCH					
SIZE OF MEMBER	ALLOWABLE MOMENT (1),(3) (KIP-FT)	SIZE OF MEMBER	ALLOWABLE MOMENT (1),(3) (KIP-FT)	SIZE OF MEMBER	ALLOWABLE MOMENT (1),(3) (KIP-FT)
2x4	0.39	(2)-2x4's	0.79	(3)-2x4's	1.18
2x6	0.84	(2)-2x6's	1.69	(3)-2x6's	2.54
2x8	1.34	(2)-2x8's	2.71	(3)-2x8's	4.07
2x10	2.01	(2)-2x10's	4.04	(3)-2x10's	6.07
2x12	2.69	(2)-2x12's	5.44	(3)-2x12's	8.17
2x14	3.37	(2)-2x14's	6.79	(3)-2x14's	10.20

#2 DOUGLAS FIR-LARCH (NORTH)					
SIZE OF MEMBER	ALLOWABLE MOMENT (1),(3) (KIP-FT)	SIZE OF MEMBER	ALLOWABLE MOMENT (1),(3) (KIP-FT)	SIZE OF MEMBER	ALLOWABLE MOMENT (1),(3) (KIP-FT)
2x4	0.37	(2)-2x4's	0.74	(3)-2x4's	1.12
2x6	0.79	(2)-2x6's	1.59	(3)-2x6's	2.40
2x8	1.27	(2)-2x8's	2.56	(3)-2x8's	3.85
2x10	1.89	(2)-2x10's	3.82	(3)-2x10's	5.74
2x12	2.55	(2)-2x12's	5.14	(3)-2x12's	7.71
2x14	3.18	(2)-2x14's	6.42	(3)-2x14's	9.64

NOTE (1) : FACTORS USED CD = 1.0, Cr = 1.15
NOTE (2): N.A. INDICATES THAT SIZE OF LUMBER IS NOT COMMONLY AVAILABLE

NOTE (3): MEMBER OWN WEIGHT HAS BEEN FACTORED INTO THE CALCULATIONS

NOTE (4): #2 WOOD IS THE MOST COMMONLY AVAILABLE GRADE COMMERCIALLY

TABLE 7-7

SAWN CUT LUMBER BEAMS / GIRDERS STRUCTURAL CAPACITY					
#2 EASTERN HEMLOCK-BALSAM FIR					
SIZE OF MEMBER	ALLOWABLE MOMENT (1),(3) (KIP-FT)	SIZE OF MEMBER	ALLOWABLE MOMENT (1),(3) (KIP-FT)	SIZE OF MEMBER	ALLOWABLE MOMENT (1),(3) (KIP-FT)
2x4	0.25	(2)-2x4's	0.50	(3)-2x4's	0.75
2x6	0.53	(2)-2x6's	1.08	(3)-2x6's	1.62
2x8	0.86	(2)-2x8's	1.73	(3)-2x8's	2.60
2x10	1.28	(2)-2x10's	2.58	(3)-2x10's	3.88
2x12	1.72	(2)-2x12's	3.48	(3)-2x12's	5.22
2x14	2.15	(2)-2x14's	4.34	(3)-2x14's	6.51

#2 EASTERN HEMLOCK-TAMARACK					
SIZE OF MEMBER	ALLOWABLE MOMENT (1),(3) (KIP-FT)	SIZE OF MEMBER	ALLOWABLE MOMENT (1),(3) (KIP-FT)	SIZE OF MEMBER	ALLOWABLE MOMENT (1),(3) (KIP-FT)
2x4	0.25	(2)-2x4's	0.50	(3)-2x4's	0.75
2x6	0.53	(2)-2x6's	1.08	(3)-2x6's	1.62
2x8	0.86	(2)-2x8's	1.73	(3)-2x8's	2.60
2x10	1.28	(2)-2x10's	2.58	(3)-2x10's	3.88
2x12	1.72	(2)-2x12's	3.48	(3)-2x12's	5.22
2x14	2.15	(2)-2x14's	4.34	(3)-2x14's	6.51

NOTE (1) : FACTORS USED CD = 1.0, Cr = 1.15
NOTE (2): N.A. INDICATES THAT SIZE OF LUMBER IS NOT COMMONLY AVAILABLE

NOTE (3): MEMBER OWN WEIGHT HAS BEEN FACTORED INTO THE CALCULATIONS

NOTE (4): #2 WOOD IS THE MOST COMMONLY AVAILABLE GRADE COMMERCIALLY

TABLE 7-8

SAWN CUT LUMBER BEAMS / GIRDERS STRUCTURAL CAPACITY					
#2 EASTERN SOFTWOODS					
SIZE OF MEMBER	ALLOWABLE MOMENT (1),(3) (KIP-FT)	SIZE OF MEMBER	ALLOWABLE MOMENT (1),(3) (KIP-FT)	SIZE OF MEMBER	ALLOWABLE MOMENT (1),(3) (KIP-FT)
2x4	0.25	(2)-2x4's	0.50	(3)-2x4's	0.75
2x6	0.53	(2)-2x6's	1.08	(3)-2x6's	1.62
2x8	0.86	(2)-2x8's	1.73	(3)-2x8's	2.60
2x10	1.28	(2)-2x10's	2.58	(3)-2x10's	3.88
2x12	1.72	(2)-2x12's	3.48	(3)-2x12's	5.22
2x14	2.15	(2)-2x14's	4.34	(3)-2x14's	6.51

#2 EASTERN WHITE PINE					
SIZE OF MEMBER	ALLOWABLE MOMENT (1),(3) (KIP-FT)	SIZE OF MEMBER	ALLOWABLE MOMENT (1),(3) (KIP-FT)	SIZE OF MEMBER	ALLOWABLE MOMENT (1),(3) (KIP-FT)
2x4	0.25	(2)-2x4's	0.50	(3)-2x4's	0.75
2x6	0.53	(2)-2x6's	1.08	(3)-2x6's	1.62
2x8	0.86	(2)-2x8's	1.73	(3)-2x8's	2.60
2x10	1.28	(2)-2x10's	2.58	(3)-2x10's	3.88
2x12	1.72	(2)-2x12's	3.48	(3)-2x12's	5.22
2x14	2.15	(2)-2x14's	4.34	(3)-2x14's	6.51

NOTE (1) : FACTORS USED CD = 1.0, Cr = 1.15
NOTE (2): N.A. INDICATES THAT SIZE OF LUMBER IS NOT COMMONLY AVAILABLE

NOTE (3): MEMBER OWN WEIGHT HAS BEEN FACTORED INTO THE CALCULATIONS

NOTE (4): #2 WOOD IS THE MOST COMMONLY AVAILABLE GRADE COMMERCIALLY

TABLE 7-9

SAWN CUT LUMBER BEAMS / GIRDERS STRUCTURAL CAPACITY					
#2 HEM-FIR					
SIZE OF MEMBER	ALLOWABLE MOMENT (1),(3) (KIP-FT)	SIZE OF MEMBER	ALLOWABLE MOMENT (1),(3) (KIP-FT)	SIZE OF MEMBER	ALLOWABLE MOMENT (1),(3) (KIP-FT)
2x4	0.37	(2)-2x4's	0.74	(3)-2x4's	1.12
2x6	0.79	(2)-2x6's	1.59	(3)-2x6's	2.4
2x8	1.27	(2)-2x8's	2.56	(3)-2x8's	3.85
2x10	1.89	(2)-2x10's	3.82	(3)-2x10's	5.74
2x12	2.54	(2)-2x12's	5.13	(3)-2x12's	7.71
2x14	3.17	(2)-2x14's	6.42	(3)-2x14's	9.64

#2 HEM-FIR (NORTH)					
SIZE OF MEMBER	ALLOWABLE MOMENT (1),(3) (KIP-FT)	SIZE OF MEMBER	ALLOWABLE MOMENT (1),(3) (KIP-FT)	SIZE OF MEMBER	ALLOWABLE MOMENT (1),(3) (KIP-FT)
2x4	0.43	(2)-2x4's	0.87	(3)-2x4's	1.31
2x6	0.93	(2)-2x6's	1.88	(3)-2x6's	2.82
2x8	1.49	(2)-2x8's	3.01	(3)-2x8's	4.53
2x10	2.23	(2)-2x10's	4.49	(3)-2x10's	6.76
2x12	2.99	(2)-2x12's	6.04	(3)-2x12's	9.08
2x14	3.73	(2)-2x14's	7.55	(3)-2x14's	11.36

NOTE (1) : FACTORS USED CD = 1.0, Cr = 1.15
NOTE (2): N.A. INDICATES THAT SIZE OF LUMBER IS NOT COMMONLY AVAILABLE

NOTE (3): MEMBER OWN WEIGHT HAS BEEN FACTORED INTO THE CALCULATIONS

NOTE (4): #2 WOOD IS THE MOST COMMONLY AVAILABLE GRADE COMMERCIALLY

TABLE 7-10

SAWN CUT LUMBER BEAMS / GIRDERS STRUCTURAL CAPACITY					
#2 MIXED MAPLE					
SIZE OF MEMBER	ALLOWABLE MOMENT (1),(3) (KIP-FT)	SIZE OF MEMBER	ALLOWABLE MOMENT (1),(3) (KIP-FT)	SIZE OF MEMBER	ALLOWABLE MOMENT (1),(3) (KIP-FT)
2x4	0.30	(2)-2x4's	0.61	(3)-2x4's	0.92
2x6	0.65	(2)-2x6's	1.31	(3)-2x6's	1.97
2x8	1.04	(2)-2x8's	2.11	(3)-2x8's	3.16
2x10	1.56	(2)-2x10's	3.14	(3)-2x10's	4.73
2x12	2.09	(2)-2x12's	4.23	(3)-2x12's	6.36
2x14	2.61	(2)-2x14's	5.28	(3)-2x14's	7.93

#2 MIXED OAK					
SIZE OF MEMBER	ALLOWABLE MOMENT (1),(3) (KIP-FT)	SIZE OF MEMBER	ALLOWABLE MOMENT (1),(3) (KIP-FT)	SIZE OF MEMBER	ALLOWABLE MOMENT (1),(3) (KIP-FT)
2x4	0.34	(2)-2x4's	0.70	(3)-2x4's	1.05
2x6	0.74	(2)-2x6's	1.50	(3)-2x6's	2.25
2x8	1.19	(2)-2x8's	2.40	(3)-2x8's	3.62
2x10	1.77	(2)-2x10's	3.59	(3)-2x10's	5.40
2x12	2.37	(2)-2x12's	4.83	(3)-2x12's	7.26
2x14	2.96	(2)-2x14's	6.03	(3)-2x14's	9.06

NOTE (1) : FACTORS USED CD = 1.0, Cr = 1.15
NOTE (2): N.A. INDICATES THAT SIZE OF LUMBER IS NOT COMMONLY AVAILABLE

NOTE (3): MEMBER OWN WEIGHT HAS BEEN FACTORED INTO THE CALCULATIONS

NOTE (4): #2 WOOD IS THE MOST COMMONLY AVAILABLE GRADE COMMERCIALLY

TABLE 7-11

SAWN CUT LUMBER BEAMS / GIRDERS STRUCTURAL CAPACITY					
#2 MIXED SOUTHERN PINE					
SIZE OF MEMBER	ALLOWABLE MOMENT [(1),(3)] (KIP-FT)	SIZE OF MEMBER	ALLOWABLE MOMENT [(1),(3)] (KIP-FT)	SIZE OF MEMBER	ALLOWABLE MOMENT [(1),(3)] (KIP-FT)
2x4	0.32	(2)-2x4's	0.64	(3)-2x4's	0.96
2x6	0.72	(2)-2x6's	1.44	(3)-2x6's	2.17
2x8	1.15	(2)-2x8's	2.32	(3)-2x8's	3.49
2x10	1.62	(2)-2x10's	3.27	(3)-2x10's	4.91
2x12	2.25	(2)-2x12's	4.53	(3)-2x12's	6.81
2x14	N.A. [(2)]	(2)-2x14's	N.A. [(2)]	(3)-2x14's	N.A. [(2)]

#2 NORTHERN RED OAK					
SIZE OF MEMBER	ALLOWABLE MOMENT [(1),(3)] (KIP-FT)	SIZE OF MEMBER	ALLOWABLE MOMENT [(1),(3)] (KIP-FT)	SIZE OF MEMBER	ALLOWABLE MOMENT [(1),(3)] (KIP-FT)
2x4	0.42	(2)-2x4's	0.85	(3)-2x4's	1.28
2x6	0.91	(2)-2x6's	1.83	(3)-2x6's	2.75
2x8	1.45	(2)-2x8's	2.94	(3)-2x8's	4.41
2x10	2.16	(2)-2x10's	4.38	(3)-2x10's	6.58
2x12	2.91	(2)-2x12's	5.89	(3)-2x12's	8.85
2x14	3.63	(2)-2x14's	7.35	(3)-2x14's	11.06

NOTE (1) : FACTORS USED CD = 1.0, Cr = 1.15
NOTE (2): N.A. INDICATES THAT SIZE OF LUMBER IS NOT COMMONLY AVAILABLE

NOTE (3): MEMBER OWN WEIGHT HAS BEEN FACTORED INTO THE CALCULATIONS

NOTE (4): #2 WOOD IS THE MOST COMMONLY AVAILABLE GRADE COMMERCIALLY

TABLE 7-12

SAWN CUT LUMBER BEAMS / GIRDERS STRUCTURAL CAPACITY					
#2 NORTHERN SPECIES					
SIZE OF MEMBER	ALLOWABLE MOMENT (1),(3) (KIP-FT)	SIZE OF MEMBER	ALLOWABLE MOMENT (1),(3) (KIP-FT)	SIZE OF MEMBER	ALLOWABLE MOMENT (1),(3) (KIP-FT)
2x4	0.27	(2)-2x4's	0.54	(3)-2x4's	0.82
2x6	0.58	(2)-2x6's	1.17	(3)-2x6's	1.76
2x8	0.93	(2)-2x8's	1.88	(3)-2x8's	2.83
2x10	1.39	(2)-2x10's	2.81	(3)-2x10's	4.22
2x12	1.87	(2)-2x12's	3.78	(3)-2x12's	5.67
2x14	2.33	(2)-2x14's	4.71	(3)-2x14's	7.08
#2 NORTHERN WHITE CEDAR					
SIZE OF MEMBER	ALLOWABLE MOMENT (1),(3) (KIP-FT)	SIZE OF MEMBER	ALLOWABLE MOMENT (1),(3) (KIP-FT)	SIZE OF MEMBER	ALLOWABLE MOMENT (1),(3) (KIP-FT)
2x4	0.24	(2)-2x4's	0.48	(3)-2x4's	0.72
2x6	0.51	(2)-2x6's	1.03	(3)-2x6's	1.55
2x8	0.82	(2)-2x8's	1.65	(3)-2x8's	2.49
2x10	1.22	(2)-2x10's	2.47	(3)-2x10's	3.71
2x12	1.64	(2)-2x12's	3.32	(3)-2x12's	4.99
2x14	2.04	(2)-2x14's	4.15	(3)-2x14's	6.23

NOTE (1) : FACTORS USED CD = 1.0, Cr = 1.15
NOTE (2): N.A. INDICATES THAT SIZE OF LUMBER IS NOT COMMONLY AVAILABLE

NOTE (3): MEMBER OWN WEIGHT HAS BEEN FACTORED INTO THE CALCULATIONS

NOTE (4): #2 WOOD IS THE MOST COMMONLY AVAILABLE GRADE COMMERCIALLY

TABLE 7-13

SAWN CUT LUMBER BEAMS / GIRDERS STRUCTURAL CAPACITY					
#2 RED MAPLE					
SIZE OF MEMBER	ALLOWABLE MOMENT (1),(3) (KIP-FT)	SIZE OF MEMBER	ALLOWABLE MOMENT (1),(3) (KIP-FT)	SIZE OF MEMBER	ALLOWABLE MOMENT (1),(3) (KIP-FT)
2x4	0.39	(2)-2x4's	0.79	(3)-2x4's	1.18
2x6	0.84	(2)-2x6's	1.69	(3)-2x6's	2.54
2x8	1.34	(2)-2x8's	2.71	(3)-2x8's	4.07
2x10	2.00	(2)-2x10's	4.05	(3)-2x10's	6.08
2x12	2.69	(2)-2x12's	5.44	(3)-2x12's	8.17
2x14	3.36	(2)-2x14's	6.80	(3)-2x14's	10.20
#2 RED OAK					
SIZE OF MEMBER	ALLOWABLE MOMENT (1),(3) (KIP-FT)	SIZE OF MEMBER	ALLOWABLE MOMENT (1),(3) (KIP-FT)	SIZE OF MEMBER	ALLOWABLE MOMENT (1),(3) (KIP-FT)
2x4	0.35	(2)-2x4's	0.70	(3)-2x4's	1.05
2x6	0.74	(2)-2x6's	1.50	(3)-2x6's	2.26
2x8	1.19	(2)-2x8's	2.41	(3)-2x8's	3.62
2x10	1.78	(2)-2x10's	3.60	(3)-2x10's	5.40
2x12	2.39	(2)-2x12's	4.84	(3)-2x12's	7.27
2x14	2.98	(2)-2x14's	6.04	(3)-2x14's	9.06

NOTE (1) : FACTORS USED CD = 1.0, Cr = 1.15
NOTE (2): N.A. INDICATES THAT SIZE OF LUMBER IS NOT COMMONLY AVAILABLE

NOTE (3): MEMBER OWN WEIGHT HAS BEEN FACTORED INTO THE CALCULATIONS

NOTE (4): #2 WOOD IS THE MOST COMMONLY AVAILABLE GRADE COMMERCIALLY

TABLE 7-14

SAWN CUT LUMBER BEAMS / GIRDERS STRUCTURAL CAPACITY					
#2 REDWOOD					
SIZE OF MEMBER	ALLOWABLE MOMENT (1),(3) (KIP-FT)	SIZE OF MEMBER	ALLOWABLE MOMENT (1),(3) (KIP-FT)	SIZE OF MEMBER	ALLOWABLE MOMENT (1),(3) (KIP-FT)
2x4	0.40	(2)-2x4's	0.81	(3)-2x4's	1.22
2x6	0.86	(2)-2x6's	1.73	(3)-2x6's	2.61
2x8	1.38	(2)-2x8's	2.78	(3)-2x8's	4.18
2x10	2.05	(2)-2x10's	4.15	(3)-2x10's	6.25
2x12	2.76	(2)-2x12's	5.59	(3)-2x12's	8.39
2x14	3.43	(2)-2x14's	6.97	(3)-2x14's	10.47
#2 SOUTHERN PINE					
SIZE OF MEMBER	ALLOWABLE MOMENT (1),(3) (KIP-FT)	SIZE OF MEMBER	ALLOWABLE MOMENT (1),(3) (KIP-FT)	SIZE OF MEMBER	ALLOWABLE MOMENT (1),(3) (KIP-FT)
2x4	0.32	(2)-2x4's	0.64	(3)-2x4's	0.96
2x6	0.72	(2)-2x6's	1.44	(3)-2x6's	2.17
2x8	1.15	(2)-2x8's	2.32	(3)-2x8's	3.49
2x10	1.62	(2)-2x10's	3.27	(3)-2x10's	4.91
2x12	2.25	(2)-2x12's	4.53	(3)-2x12's	6.81
2x14	N.A. (2)	(2)-2x14's	N.A. (2)	(3)-2x14's	N.A. (2)

NOTE (1) : FACTORS USED CD = 1.0, Cr = 1.15
NOTE (2): N.A. INDICATES THAT SIZE OF LUMBER IS NOT COMMONLY AVAILABLE

NOTE (3): MEMBER OWN WEIGHT HAS BEEN FACTORED INTO THE CALCULATIONS

NOTE (4): #2 WOOD IS THE MOST COMMONLY AVAILABLE GRADE COMMERCIALLY

TABLE 7-15

SAWN CUT LUMBER BEAMS / GIRDERS STRUCTURAL CAPACITY					
#2 SPRUCE-PINE-FIR (SOUTH)					
SIZE OF MEMBER	ALLOWABLE MOMENT (1),(3) (KIP-FT)	SIZE OF MEMBER	ALLOWABLE MOMENT (1),(3) (KIP-FT)	SIZE OF MEMBER	ALLOWABLE MOMENT (1),(3) (KIP-FT)
2x4	0.33	(2)-2x4's	0.68	(3)-2x4's	1.02
2x6	0.72	(2)-2x6's	1.45	(3)-2x6's	2.18
2x8	1.15	(2)-2x8's	2.33	(3)-2x8's	3.51
2x10	1.72	(2)-2x10's	3.48	(3)-2x10's	5.23
2x12	2.31	(2)-2x12's	4.68	(3)-2x12's	7.04
2x14	2.88	(2)-2x14's	5.85	(3)-2x14's	8.80

#2 SPRUCE-PINE-FIR					
SIZE OF MEMBER	ALLOWABLE MOMENT (1),(3) (KIP-FT)	SIZE OF MEMBER	ALLOWABLE MOMENT (1),(3) (KIP-FT)	SIZE OF MEMBER	ALLOWABLE MOMENT (1),(3) (KIP-FT)
2x4	0.38	(2)-2x4's	0.76	(3)-2x4's	1.15
2x6	0.81	(2)-2x6's	1.64	(3)-2x6's	2.47
2x8	1.31	(2)-2x8's	2.63	(3)-2x8's	3.96
2x10	1.95	(2)-2x10's	3.93	(3)-2x10's	5.91
2x12	2.62	(2)-2x12's	5.29	(3)-2x12's	7.95
2x14	3.27	(2)-2x14's	6.60	(3)-2x14's	9.93

NOTE (1) : FACTORS USED CD = 1.0, Cr = 1.15
NOTE (2): N.A. INDICATES THAT SIZE OF LUMBER IS NOT COMMONLY AVAILABLE

NOTE (3): MEMBER OWN WEIGHT HAS BEEN FACTORED INTO THE CALCULATIONS

NOTE (4): #2 WOOD IS THE MOST COMMONLY AVAILABLE GRADE COMMERCIALLY

TABLE 7-16

SAWN CUT LUMBER BEAMS / GIRDERS STRUCTURAL CAPACITY					
#2 WESTERN CEDARS					
SIZE OF MEMBER	ALLOWABLE MOMENT (1),(3) (KIP-FT)	SIZE OF MEMBER	ALLOWABLE MOMENT (1),(3) (KIP-FT)	SIZE OF MEMBER	ALLOWABLE MOMENT (1),(3) (KIP-FT)
2x4	0.30	(2)-2x4's	0.61	(3)-2x4's	0.92
2x6	0.65	(2)-2x6's	1.31	(3)-2x6's	1.97
2x8	1.04	(2)-2x8's	2.11	(3)-2x8's	3.16
2x10	1.56	(2)-2x10's	3.14	(3)-2x10's	4.73
2x12	2.09	(2)-2x12's	4.23	(3)-2x12's	6.36
2x14	2.61	(2)-2x14's	5.28	(3)-2x14's	7.93
#2 WESTERN WOODS					
SIZE OF MEMBER	ALLOWABLE MOMENT (1),(3) (KIP-FT)	SIZE OF MEMBER	ALLOWABLE MOMENT (1),(3) (KIP-FT)	SIZE OF MEMBER	ALLOWABLE MOMENT (1),(3) (KIP-FT)
2x4	0.29	(2)-2x4's	0.59	(3)-2x4's	0.89
2x6	0.63	(2)-2x6's	1.26	(3)-2x6's	1.90
2x8	1.00	(2)-2x8's	2.03	(3)-2x8's	3.05
2x10	1.50	(2)-2x10's	3.03	(3)-2x10's	4.56
2x12	2.02	(2)-2x12's	4.07	(3)-2x12's	6.13
2x14	2.51	(2)-2x14's	5.09	(3)-2x14's	7.65

NOTE (1) : FACTORS USED CD = 1.0, Cr = 1.15
NOTE (2): N.A. INDICATES THAT SIZE OF LUMBER IS NOT COMMONLY AVAILABLE

NOTE (3): MEMBER OWN WEIGHT HAS BEEN FACTORED INTO THE CALCULATIONS

NOTE (4): #2 WOOD IS THE MOST COMMONLY AVAILABLE GRADE COMMERCIALLY

TABLE 7-17

SAWN CUT LUMBER BEAMS / GIRDERS STRUCTURAL CAPACITY					
#2 WHITE OAK					
SIZE OF MEMBER	ALLOWABLE MOMENT [(1),(3)] (KIP-FT)	SIZE OF MEMBER	ALLOWABLE MOMENT [(1),(3)] (KIP-FT)	SIZE OF MEMBER	ALLOWABLE MOMENT [(1),(3)] (KIP-FT)
2x4	0.37	(2)-2x4's	0.74	(3)-2x4's	1.12
2x6	0.79	(2)-2x6's	1.59	(3)-2x6's	2.4
2x8	1.27	(2)-2x8's	2.56	(3)-2x8's	3.85
2x10	1.89	(2)-2x10's	3.82	(3)-2x10's	5.74
2x12	2.54	(2)-2x12's	5.13	(3)-2x12's	7.71
2x14	3.17	(2)-2x14's	6.42	(3)-2x14's	9.64
#2 YELLOW CEDAR					
SIZE OF MEMBER	ALLOWABLE MOMENT [(1),(3)] (KIP-FT)	SIZE OF MEMBER	ALLOWABLE MOMENT [(1),(3)] (KIP-FT)	SIZE OF MEMBER	ALLOWABLE MOMENT [(1),(3)] (KIP-FT)
2x4	0.35	(2)-2x4's	0.70	(3)-2x4's	1.05
2x6	0.74	(2)-2x6's	1.50	(3)-2x6's	2.26
2x8	1.19	(2)-2x8's	2.41	(3)-2x8's	3.62
2x10	1.78	(2)-2x10's	3.60	(3)-2x10's	5.40
2x12	2.39	(2)-2x12's	4.84	(3)-2x12's	7.27
2x14	2.98	(2)-2x14's	6.04	(3)-2x14's	9.06

NOTE (1) : FACTORS USED CD = 1.0, Cr = 1.15
NOTE (2): N.A. INDICATES THAT SIZE OF LUMBER IS NOT COMMONLY AVAILABLE

NOTE (3): MEMBER OWN WEIGHT HAS BEEN FACTORED INTO THE CALCULATIONS

NOTE (4): #2 WOOD IS THE MOST COMMONLY AVAILABLE GRADE COMMERCIALLY

TABLE 7-18

SAWN CUT LUMBER BEAMS / GIRDERS STRUCTURAL CAPACITY					
#2 YELLOW POPLAR					
SIZE OF MEMBER	ALLOWABLE MOMENT (1),(3) (KIP-FT)	SIZE OF MEMBER	ALLOWABLE MOMENT (1),(3) (KIP-FT)	SIZE OF MEMBER	ALLOWABLE MOMENT (1),(3) (KIP-FT)
2x4	0.30	(2)-2x4's	0.61	(3)-2x4's	0.92
2x6	0.65	(2)-2x6's	1.31	(3)-2x6's	1.97
2x8	1.04	(2)-2x8's	2.11	(3)-2x8's	3.16
2x10	1.56	(2)-2x10's	3.14	(3)-2x10's	4.73
2x12	2.09	(2)-2x12's	4.23	(3)-2x12's	6.36
2x14	2.62	(2)-2x14's	5.29	(3)-2x14's	7.93

NOTE (1) : FACTORS USED CD = 1.0, Cr = 1.15
NOTE (2): N.A. INDICATES THAT SIZE OF LUMBER IS NOT COMMONLY AVAILABLE

NOTE (3): MEMBER OWN WEIGHT HAS BEEN FACTORED INTO THE CALCULATIONS

NOTE (4): #2 WOOD IS THE MOST COMMONLY AVAILABLE GRADE COMMERCIALLY

TABLE 7-19

SAWN CUT TIMBER POSTS STRUCTURAL CAPACITY					
#2 ALASKA CEDAR					
COLUMN CLEAR HEIGHT (FT)	4x4 POST STRUCTURAL CAPACITY (KIPS)(1),(2),(3)	6x6 POST STRUCTURAL CAPACITY (KIPS)(1),(2),(3)	8x8 POST STRUCTURAL CAPACITY (KIPS)(1),(2),(3)	10x10 POST STRUCTURAL CAPACITY (KIPS)(1),(2),(3)	12x12 POST STRUCTURAL CAPACITY (KIPS)(1),(2),(3)
4.00	9.33	17.54	33.18	53.60	78.82
6.00	7.25	16.63	32.34	52.79	78.02
8.00	4.97	15.06	31.05	51.62	76.96
10.00	3.43	12.85	29.13	49.99	75.38
12.00	2.46	10.43	26.49	47.65	73.39
14.00	1.83	8.31	23.28	44.67	70.75
16.00	-	6.68	19.96	40.97	67.31
18.00	-	5.41	16.93	36.82	63.21
20.00	-	4.47	14.34	32.58	58.42
22.00	-	3.75	12.26	28.60	53.42
24.00	-	-	10.51	25.08	48.27
26.00	-	-	9.11	22.11	43.37
28.00	-	-	7.93	19.49	39.01
30.00	-	-	6.97	17.23	35.04
32.00	-	-	-	15.34	31.47
34.00	-	-	-	13.71	28.43
36.00	-	-	-	12.36	25.65
38.00	-	-	-	11.19	23.27
40.00	-	-	-	-	21.29
42.00	-	-	-	-	19.44

NOTE (1) : FACTORS USED CD = 1.0, CF = VARIES BASED ON SIZE, CM = 1.0, Ct = 1.0, Ci = 1.0, CT = 1.0

NOTE (2): MEMBER OWN WEIGHT HAS BEEN FACTORED INTO THE CALCULATIONS

NOTE (3): #2 WOOD IS THE MOST COMMONLY AVAILABLE GRADE COMMERCIALLY

NOTE (4): N.A. INDICATES THAT SIZE OF TIMBER IS NOT COMMONLY AVAILABLE

TABLE 7-20

SAWN CUT TIMBER POSTS STRUCTURAL CAPACITY					
#2 ALASKA HEMLOCK					
COLUMN CLEAR HEIGHT (FT)	4x4 POST STRUCTURAL CAPACITY (KIPS)(1),(2),(3)	6x6 POST STRUCTURAL CAPACITY (KIPS)(1),(2),(3)	8x8 POST STRUCTURAL CAPACITY (KIPS)(1),(2),(3)	10x10 POST STRUCTURAL CAPACITY (KIPS)(1),(2),(3)	12x12 POST STRUCTURAL CAPACITY (KIPS)(1),(2),(3)
4.00	12.80	N.A.	N.A.	N.A.	N.A.
6.00	9.55	N.A.	N.A.	N.A.	N.A.
8.00	6.37	N.A.	N.A.	N.A.	N.A.
10.00	4.33	N.A.	N.A.	N.A.	N.A.
12.00	3.11	N.A.	N.A.	N.A.	N.A.
14.00	2.31	N.A.	N.A.	N.A.	N.A.
16.00	-	N.A.	N.A.	N.A.	N.A.
18.00	-	N.A.	N.A.	N.A.	N.A.
20.00	-	N.A.	N.A.	N.A.	N.A.
22.00	-	N.A.	N.A.	N.A.	N.A.
24.00	-	N.A.	N.A.	N.A.	N.A.
26.00	-	N.A.	N.A.	N.A.	N.A.
28.00	-	N.A.	N.A.	N.A.	N.A.
30.00	-	N.A.	N.A.	N.A.	N.A.
32.00	-	N.A.	N.A.	N.A.	N.A.
34.00	-	N.A.	N.A.	N.A.	N.A.
36.00	-	N.A.	N.A.	N.A.	N.A.
38.00	-	N.A.	N.A.	N.A.	N.A.
40.00	-	N.A.	N.A.	N.A.	N.A.
42.00	-	N.A.	N.A.	N.A.	N.A.

NOTE (1) : FACTORS USED CD = 1.0, CF = VARIES BASED ON SIZE, CM = 1.0, Ct = 1.0, Ci = 1.0, CT = 1.0

NOTE (2): MEMBER OWN WEIGHT HAS BEEN FACTORED INTO THE CALCULATIONS

NOTE (3): #2 WOOD IS THE MOST COMMONLY AVAILABLE GRADE COMMERCIALLY

NOTE (4): N.A. INDICATES THAT SIZE OF TIMBER IS NOT COMMONLY AVAILABLE

TABLE 7-21

SAWN CUT TIMBER POSTS STRUCTURAL CAPACITY					
#2 ALASKA SPRUCE					
COLUMN CLEAR HEIGHT (FT)	4x4 POST STRUCTURAL CAPACITY (KIPS)(1),(2),(3)	6x6 POST STRUCTURAL CAPACITY (KIPS)(1),(2),(3)	8x8 POST STRUCTURAL CAPACITY (KIPS)(1),(2),(3)	10x10 POST STRUCTURAL CAPACITY (KIPS)(1),(2),(3)	12x12 POST STRUCTURAL CAPACITY (KIPS)(1),(2),(3)
4.00	12.62	N.A.	N.A.	N.A.	N.A.
6.00	9.17	N.A.	N.A.	N.A.	N.A.
8.00	6.01	N.A.	N.A.	N.A.	N.A.
10.00	4.06	N.A.	N.A.	N.A.	N.A.
12.00	2.90	N.A.	N.A.	N.A.	N.A.
14.00	2.15	N.A.	N.A.	N.A.	N.A.
16.00	-	N.A.	N.A.	N.A.	N.A.
18.00	-	N.A.	N.A.	N.A.	N.A.
20.00	-	N.A.	N.A.	N.A.	N.A.
22.00	-	N.A.	N.A.	N.A.	N.A.
24.00	-	N.A.	N.A.	N.A.	N.A.
26.00	-	N.A.	N.A.	N.A.	N.A.
28.00	-	N.A.	N.A.	N.A.	N.A.
30.00	-	N.A.	N.A.	N.A.	N.A.
32.00	-	N.A.	N.A.	N.A.	N.A.
34.00	-	N.A.	N.A.	N.A.	N.A.
36.00	-	N.A.	N.A.	N.A.	N.A.
38.00	-	N.A.	N.A.	N.A.	N.A.
40.00	-	N.A.	N.A.	N.A.	N.A.
42.00	-	N.A.	N.A.	N.A.	N.A.

NOTE (1) : FACTORS USED CD = 1.0, CF = VARIES BASED ON SIZE, CM = 1.0, Ct = 1.0, Ci = 1.0, CT = 1.0

NOTE (2): MEMBER OWN WEIGHT HAS BEEN FACTORED INTO THE CALCULATIONS

NOTE (3): #2 WOOD IS THE MOST COMMONLY AVAILABLE GRADE COMMERCIALLY

NOTE (4): N.A. INDICATES THAT SIZE OF TIMBER IS NOT COMMONLY AVAILABLE

TABLE 7-22

SAWN CUT TIMBER POSTS STRUCTURAL CAPACITY					
#2 ALASKA YELLOW CEDAR					
COLUMN CLEAR HEIGHT (FT)	4x4 POST STRUCTURAL CAPACITY (KIPS)(1),(2),(3)	6x6 POST STRUCTURAL CAPACITY (KIPS)(1),(2),(3)	8x8 POST STRUCTURAL CAPACITY (KIPS)(1),(2),(3)	10x10 POST STRUCTURAL CAPACITY (KIPS)(1),(2),(3)	12x12 POST STRUCTURAL CAPACITY (KIPS)(1),(2),(3)
4.00	11.95	N.A.	N.A.	N.A.	N.A.
6.00	8.59	N.A.	N.A.	N.A.	N.A.
8.00	5.61	N.A.	N.A.	N.A.	N.A.
10.00	3.78	N.A.	N.A.	N.A.	N.A.
12.00	2.69	N.A.	N.A.	N.A.	N.A.
14.00	2.00	N.A.	N.A.	N.A.	N.A.
16.00	-	N.A.	N.A.	N.A.	N.A.
18.00	-	N.A.	N.A.	N.A.	N.A.
20.00	-	N.A.	N.A.	N.A.	N.A.
22.00	-	N.A.	N.A.	N.A.	N.A.
24.00	-	N.A.	N.A.	N.A.	N.A.
26.00	-	N.A.	N.A.	N.A.	N.A.
28.00	-	N.A.	N.A.	N.A.	N.A.
30.00	-	N.A.	N.A.	N.A.	N.A.
32.00	-	N.A.	N.A.	N.A.	N.A.
34.00	-	N.A.	N.A.	N.A.	N.A.
36.00	-	N.A.	N.A.	N.A.	N.A.
38.00	-	N.A.	N.A.	N.A.	N.A.
40.00	-	N.A.	N.A.	N.A.	N.A.
42.00	-	N.A.	N.A.	N.A.	N.A.

NOTE (1) : FACTORS USED CD = 1.0, CF = VARIES BASED ON SIZE, CM = 1.0, Ct = 1.0, Ci = 1.0, CT = 1.0

NOTE (2): MEMBER OWN WEIGHT HAS BEEN FACTORED INTO THE CALCULATIONS

NOTE (3): #2 WOOD IS THE MOST COMMONLY AVAILABLE GRADE COMMERCIALLY

NOTE (4): N.A. INDICATES THAT SIZE OF TIMBER IS NOT COMMONLY AVAILABLE

TABLE 7-23

SAWN CUT TIMBER POSTS STRUCTURAL CAPACITY					
#2 ASPEN					
COLUMN CLEAR HEIGHT (FT)	4x4 POST STRUCTURAL CAPACITY (KIPS)(1),(2),(3)	6x6 POST STRUCTURAL CAPACITY (KIPS)(1),(2),(3)	8x8 POST STRUCTURAL CAPACITY (KIPS)(1),(2),(3)	10x10 POST STRUCTURAL CAPACITY (KIPS)(1),(2),(3)	12x12 POST STRUCTURAL CAPACITY (KIPS)(1),(2),(3)
4.00	5.84	N.A.	N.A.	N.A.	N.A.
6.00	4.99	N.A.	N.A.	N.A.	N.A.
8.00	3.77	N.A.	N.A.	N.A.	N.A.
10.00	2.71	N.A.	N.A.	N.A.	N.A.
12.00	1.99	N.A.	N.A.	N.A.	N.A.
14.00	1.50	N.A.	N.A.	N.A.	N.A.
16.00	-	N.A.	N.A.	N.A.	N.A.
18.00	-	N.A.	N.A.	N.A.	N.A.
20.00	-	N.A.	N.A.	N.A.	N.A.
22.00	-	N.A.	N.A.	N.A.	N.A.
24.00	-	N.A.	N.A.	N.A.	N.A.
26.00	-	N.A.	N.A.	N.A.	N.A.
28.00	-	N.A.	N.A.	N.A.	N.A.
30.00	-	N.A.	N.A.	N.A.	N.A.
32.00	-	N.A.	N.A.	N.A.	N.A.
34.00	-	N.A.	N.A.	N.A.	N.A.
36.00	-	N.A.	N.A.	N.A.	N.A.
38.00	-	N.A.	N.A.	N.A.	N.A.
40.00	-	N.A.	N.A.	N.A.	N.A.
42.00	-	N.A.	N.A.	N.A.	N.A.

NOTE (1) : FACTORS USED CD = 1.0, CF = VARIES BASED ON SIZE, CM = 1.0, Ct = 1.0, Ci = 1.0, CT = 1.0

NOTE (2): MEMBER OWN WEIGHT HAS BEEN FACTORED INTO THE CALCULATIONS

NOTE (3): #2 WOOD IS THE MOST COMMONLY AVAILABLE GRADE COMMERCIALLY

NOTE (4): N.A. INDICATES THAT SIZE OF TIMBER IS NOT COMMONLY AVAILABLE

TABLE 7-24

SAWN CUT TIMBER POSTS STRUCTURAL CAPACITY					
#2 BALDCYPRESS					
COLUMN CLEAR HEIGHT (FT)	4x4 POST STRUCTURAL CAPACITY (KIPS)(1),(2),(3)	6x6 POST STRUCTURAL CAPACITY (KIPS)(1),(2),(3)	8x8 POST STRUCTURAL CAPACITY (KIPS)(1),(2),(3)	10x10 POST STRUCTURAL CAPACITY (KIPS)(1),(2),(3)	12x12 POST STRUCTURAL CAPACITY (KIPS)(1),(2),(3)
4.00	11.00	17.54	33.18	53.60	78.82
6.00	8.24	16.63	32.34	52.79	78.02
8.00	5.51	15.06	31.05	51.62	76.96
10.00	3.74	12.85	29.13	49.99	75.38
12.00	2.68	10.43	26.49	47.65	73.39
14.00	1.99	8.31	23.28	44.67	70.75
16.00	-	6.68	19.96	40.97	67.31
18.00	-	5.41	16.93	36.82	63.21
20.00	-	4.47	14.34	32.58	58.42
22.00	-	3.75	12.26	28.60	53.42
24.00	-	-	10.51	25.08	48.27
26.00	-	-	9.11	22.11	43.37
28.00	-	-	7.93	19.49	39.01
30.00	-	-	6.97	17.23	35.04
32.00	-	-	-	15.34	31.47
34.00	-	-	-	13.71	28.43
36.00	-	-	-	12.36	25.65
38.00	-	-	-	11.19	23.27
40.00	-	-	-	-	21.29
42.00	-	-	-	-	19.44

NOTE (1) : FACTORS USED CD = 1.0, CF = VARIES BASED ON SIZE, CM = 1.0, Ct = 1.0, Ci = 1.0, CT = 1.0

NOTE (2): MEMBER OWN WEIGHT HAS BEEN FACTORED INTO THE CALCULATIONS

NOTE (3): #2 WOOD IS THE MOST COMMONLY AVAILABLE GRADE COMMERCIALLY

NOTE (4): N.A. INDICATES THAT SIZE OF TIMBER IS NOT COMMONLY AVAILABLE

TABLE 7-25

SAWN CUT TIMBER POSTS STRUCTURAL CAPACITY					
#2 BEECH-BIRCH-HICKORY					
COLUMN CLEAR HEIGHT (FT)	4x4 POST STRUCTURAL CAPACITY (KIPS)(1),(2),(3)	6x6 POST STRUCTURAL CAPACITY (KIPS)(1),(2),(3)	8x8 POST STRUCTURAL CAPACITY (KIPS)(1),(2),(3)	10x10 POST STRUCTURAL CAPACITY (KIPS)(1),(2),(3)	12x12 POST STRUCTURAL CAPACITY (KIPS)(1),(2),(3)
4.00	9.62	12.61	23.68	38.08	55.94
6.00	8.01	12.28	23.34	37.81	55.67
8.00	5.86	11.70	22.89	37.36	55.28
10.00	4.15	10.89	22.16	36.73	54.61
12.00	3.01	9.77	21.20	35.91	53.82
14.00	2.26	8.43	19.96	34.83	52.90
16.00	-	7.13	18.45	33.48	51.70
18.00	-	6.01	16.70	31.85	50.25
20.00	-	5.05	14.90	29.87	48.53
22.00	-	4.29	13.16	27.70	46.41
24.00	-	-	11.58	25.45	44.03
26.00	-	-	10.18	23.10	41.39
28.00	-	-	9.00	20.93	38.74
30.00	-	-	7.98	18.95	35.97
32.00	-	-	-	17.14	33.19
34.00	-	-	-	15.52	30.54
36.00	-	-	-	14.07	28.16
38.00	-	-	-	12.81	25.92
40.00	-	-	-	-	23.80
42.00	-	-	-	-	21.95

NOTE (1) : FACTORS USED CD = 1.0, CF = VARIES BASED ON SIZE, CM = 1.0, Ct = 1.0, Ci = 1.0, CT = 1.0

NOTE (2): MEMBER OWN WEIGHT HAS BEEN FACTORED INTO THE CALCULATIONS

NOTE (3): #2 WOOD IS THE MOST COMMONLY AVAILABLE GRADE COMMERCIALLY

NOTE (4): N.A. INDICATES THAT SIZE OF TIMBER IS NOT COMMONLY AVAILABLE

TABLE 7-26

SAWN CUT TIMBER POSTS STRUCTURAL CAPACITY					
#2 COAST SITKA SPRUCE					
COLUMN CLEAR HEIGHT (FT)	4x4 POST STRUCTURAL CAPACITY (KIPS)(1),(2),(3)	6x6 POST STRUCTURAL CAPACITY (KIPS)(1),(2),(3)	8x8 POST STRUCTURAL CAPACITY (KIPS)(1),(2),(3)	10x10 POST STRUCTURAL CAPACITY (KIPS)(1),(2),(3)	12x12 POST STRUCTURAL CAPACITY (KIPS)(1),(2),(3)
4.00	13.29	14.79	27.78	44.76	65.86
6.00	9.73	14.30	27.33	44.40	65.33
8.00	6.41	13.49	26.66	43.77	64.80
10.00	4.34	12.28	25.65	42.86	63.87
12.00	3.09	10.70	24.30	41.69	62.81
14.00	2.31	9.01	22.50	40.07	61.49
16.00	-	7.47	20.30	38.08	59.77
18.00	-	6.20	18.00	35.73	57.52
20.00	-	5.17	15.75	33.03	55.01
22.00	-	4.38	13.72	30.05	51.97
24.00	-	-	11.98	27.16	48.66
26.00	-	-	10.46	24.36	45.22
28.00	-	-	9.22	21.84	41.65
30.00	-	-	8.15	19.58	38.22
32.00	-	-	-	17.59	34.91
34.00	-	-	-	15.88	31.87
36.00	-	-	-	14.34	29.22
38.00	-	-	-	13.08	26.71
40.00	-	-	-	-	24.46
42.00	-	-	-	-	22.48

NOTE (1) : FACTORS USED CD = 1.0, CF = VARIES BASED ON SIZE, CM = 1.0, Ct = 1.0, Ci = 1.0, CT = 1.0

NOTE (2): MEMBER OWN WEIGHT HAS BEEN FACTORED INTO THE CALCULATIONS

NOTE (3): #2 WOOD IS THE MOST COMMONLY AVAILABLE GRADE COMMERCIALLY

NOTE (4): N.A. INDICATES THAT SIZE OF TIMBER IS NOT COMMONLY AVAILABLE

TABLE 7-27

SAWN CUT TIMBER POSTS STRUCTURAL CAPACITY					
#2 COTTONWOOD					
COLUMN CLEAR HEIGHT (FT)	4x4 POST STRUCTURAL CAPACITY (KIPS)(1),(2),(3)	6x6 POST STRUCTURAL CAPACITY (KIPS)(1),(2),(3)	8x8 POST STRUCTURAL CAPACITY (KIPS)(1),(2),(3)	10x10 POST STRUCTURAL CAPACITY (KIPS)(1),(2),(3)	12x12 POST STRUCTURAL CAPACITY (KIPS)(1),(2),(3)
4.00	6.18	N.A.	N.A.	N.A.	N.A.
6.00	5.34	N.A.	N.A.	N.A.	N.A.
8.00	4.09	N.A.	N.A.	N.A.	N.A.
10.00	2.96	N.A.	N.A.	N.A.	N.A.
12.00	2.18	N.A.	N.A.	N.A.	N.A.
14.00	1.65	N.A.	N.A.	N.A.	N.A.
16.00	-	N.A.	N.A.	N.A.	N.A.
18.00	-	N.A.	N.A.	N.A.	N.A.
20.00	-	N.A.	N.A.	N.A.	N.A.
22.00	-	N.A.	N.A.	N.A.	N.A.
24.00	-	N.A.	N.A.	N.A.	N.A.
26.00	-	N.A.	N.A.	N.A.	N.A.
28.00	-	N.A.	N.A.	N.A.	N.A.
30.00	-	N.A.	N.A.	N.A.	N.A.
32.00	-	N.A.	N.A.	N.A.	N.A.
34.00	-	N.A.	N.A.	N.A.	N.A.
36.00	-	N.A.	N.A.	N.A.	N.A.
38.00	-	N.A.	N.A.	N.A.	N.A.
40.00	-	N.A.	N.A.	N.A.	N.A.
42.00	-	N.A.	N.A.	N.A.	N.A.

NOTE (1) : FACTORS USED CD = 1.0, CF = VARIES BASED ON SIZE, CM = 1.0, Ct = 1.0, Ci = 1.0, CT = 1.0

NOTE (2): MEMBER OWN WEIGHT HAS BEEN FACTORED INTO THE CALCULATIONS

NOTE (3): #2 WOOD IS THE MOST COMMONLY AVAILABLE GRADE COMMERCIALLY

NOTE (4): N.A. INDICATES THAT SIZE OF TIMBER IS NOT COMMONLY AVAILABLE

TABLE 7-28

SAWN CUT TIMBER POSTS STRUCTURAL CAPACITY					
#2 DOUGLAS FIR-SOUTH					
COLUMN CLEAR HEIGHT (FT)	4x4 POST STRUCTURAL CAPACITY (KIPS)(1),(2),(3)	6x6 POST STRUCTURAL CAPACITY (KIPS)(1),(2),(3)	8x8 POST STRUCTURAL CAPACITY (KIPS)(1),(2),(3)	10x10 POST STRUCTURAL CAPACITY (KIPS)(1),(2),(3)	12x12 POST STRUCTURAL CAPACITY (KIPS)(1),(2),(3)
4.00	14.41	18.93	35.88	58.03	85.30
6.00	8.88	17.84	34.93	57.12	84.37
8.00	5.42	15.97	33.35	55.68	83.18
10.00	3.58	13.37	31.05	53.69	81.33
12.00	2.53	10.70	27.90	50.90	78.82
14.00	1.87	8.47	24.18	47.29	75.64
16.00	-	6.74	20.53	42.95	71.54
18.00	-	5.47	17.26	38.17	66.65
20.00	-	4.50	14.56	33.48	61.09
22.00	-	3.75	12.37	29.24	55.28
24.00	-	-	10.57	25.54	49.72
26.00	-	-	9.16	22.38	44.43
28.00	-	-	7.98	19.67	39.67
30.00	-	-	6.97	17.41	35.57
32.00	-	-	-	15.43	31.87
34.00	-	-	-	13.80	28.69
36.00	-	-	-	12.45	25.92
38.00	-	-	-	11.19	23.54
40.00	-	-	-	-	21.42
42.00	-	-	-	-	19.57

NOTE (1) : FACTORS USED CD = 1.0, CF = VARIES BASED ON SIZE, CM = 1.0, Ct = 1.0, Ci = 1.0, CT = 1.0

NOTE (2): MEMBER OWN WEIGHT HAS BEEN FACTORED INTO THE CALCULATIONS

NOTE (3): #2 WOOD IS THE MOST COMMONLY AVAILABLE GRADE COMMERCIALLY

NOTE (4): N.A. INDICATES THAT SIZE OF TIMBER IS NOT COMMONLY AVAILABLE

TABLE 7-29

SAWN CUT TIMBER POSTS STRUCTURAL CAPACITY					
#2 DOUGLAS FIR-LARCH					
COLUMN CLEAR HEIGHT (FT)	4x4 POST STRUCTURAL CAPACITY (KIPS)(1),(2),(3)	6x6 POST STRUCTURAL CAPACITY (KIPS)(1),(2),(3)	8x8 POST STRUCTURAL CAPACITY (KIPS)(1),(2),(3)	10x10 POST STRUCTURAL CAPACITY (KIPS)(1),(2),(3)	12x12 POST STRUCTURAL CAPACITY (KIPS)(1),(2),(3)
4.00	15.80	20.53	38.75	58.03	92.04
6.00	10.93	19.60	37.91	57.12	91.25
8.00	7.00	18.02	36.61	55.68	90.06
10.00	4.70	15.73	34.65	53.69	88.47
12.00	3.33	13.03	31.95	50.90	86.35
14.00	2.47	10.55	28.57	47.29	83.71
16.00	-	8.53	24.91	42.95	80.27
18.00	-	6.95	21.37	38.17	76.04
20.00	-	5.77	18.28	33.48	71.15
22.00	-	4.84	15.63	29.24	65.72
24.00	-	-	13.50	25.54	60.04
26.00	-	-	11.70	22.38	54.48
28.00	-	-	10.23	19.67	49.32
30.00	-	-	9.00	17.41	44.43
32.00	-	-	-	15.43	40.20
34.00	-	-	-	13.80	36.36
36.00	-	-	-	12.45	32.93
38.00	-	-	-	11.19	30.02
40.00	-	-	-	-	27.37
42.00	-	-	-	-	24.99

NOTE (1) : FACTORS USED CD = 1.0, CF = VARIES BASED ON SIZE, CM = 1.0, Ct = 1.0, Ci = 1.0, CT = 1.0

NOTE (2): MEMBER OWN WEIGHT HAS BEEN FACTORED INTO THE CALCULATIONS

NOTE (3): #2 WOOD IS THE MOST COMMONLY AVAILABLE GRADE COMMERCIALLY

NOTE (4): N.A. INDICATES THAT SIZE OF TIMBER IS NOT COMMONLY AVAILABLE

TABLE 7-30

SAWN CUT TIMBER POSTS STRUCTURAL CAPACITY					
#2 DOUGLAS FIR-LARCH (NORTH)					
COLUMN CLEAR HEIGHT (FT)	4x4 POST STRUCTURAL CAPACITY (KIPS)(1),(2),(3)	6x6 POST STRUCTURAL CAPACITY (KIPS)(1),(2),(3)	8x8 POST STRUCTURAL CAPACITY (KIPS)(1),(2),(3)	10x10 POST STRUCTURAL CAPACITY (KIPS)(1),(2),(3)	12x12 POST STRUCTURAL CAPACITY (KIPS)(1),(2),(3)
4.00	16.23	20.53	38.75	62.54	92.04
6.00	11.07	19.60	37.91	61.82	91.25
8.00	7.04	18.02	36.61	60.55	90.06
10.00	4.71	15.73	34.65	58.93	88.47
12.00	3.34	13.03	31.95	56.58	86.35
14.00	2.48	10.55	28.57	53.51	83.71
16.00	-	8.53	24.91	49.63	80.27
18.00	-	6.95	21.37	45.30	76.04
20.00	-	5.77	18.28	40.61	71.15
22.00	-	4.84	15.63	36.00	65.72
24.00	-	-	13.50	31.85	60.04
26.00	-	-	11.70	28.15	54.48
28.00	-	-	10.23	24.90	49.32
30.00	-	-	9.00	22.11	44.43
32.00	-	-	-	19.76	40.20
34.00	-	-	-	17.68	36.36
36.00	-	-	-	15.97	32.93
38.00	-	-	-	14.40	30.02
40.00	-	-	-	-	27.37
42.00	-	-	-	-	24.99

NOTE (1) : FACTORS USED CD = 1.0, CF = VARIES BASED ON SIZE, CM = 1.0, Ct = 1.0, Ci = 1.0, CT = 1.0

NOTE (2): MEMBER OWN WEIGHT HAS BEEN FACTORED INTO THE CALCULATIONS

NOTE (3): #2 WOOD IS THE MOST COMMONLY AVAILABLE GRADE COMMERCIALLY

NOTE (4): N.A. INDICATES THAT SIZE OF TIMBER IS NOT COMMONLY AVAILABLE

TABLE 7-31

SAWN CUT TIMBER POSTS STRUCTURAL CAPACITY					
#2 EASTERN HEMLOCK-BALSAM FIR					
COLUMN CLEAR HEIGHT (FT)	4x4 POST STRUCTURAL CAPACITY (KIPS)(1),(2),(3)	6x6 POST STRUCTURAL CAPACITY (KIPS)(1),(2),(3)	8x8 POST STRUCTURAL CAPACITY (KIPS)(1),(2),(3)	10x10 POST STRUCTURAL CAPACITY (KIPS)(1),(2),(3)	12x12 POST STRUCTURAL CAPACITY (KIPS)(1),(2),(3)
4.00	9.92	N.A.	N.A.	N.A.	N.A.
6.00	7.20	N.A.	N.A.	N.A.	N.A.
8.00	4.72	N.A.	N.A.	N.A.	N.A.
10.00	3.19	N.A.	N.A.	N.A.	N.A.
12.00	2.27	N.A.	N.A.	N.A.	N.A.
14.00	1.70	N.A.	N.A.	N.A.	N.A.
16.00	-	N.A.	N.A.	N.A.	N.A.
18.00	-	N.A.	N.A.	N.A.	N.A.
20.00	-	N.A.	N.A.	N.A.	N.A.
22.00	-	N.A.	N.A.	N.A.	N.A.
24.00	-	N.A.	N.A.	N.A.	N.A.
26.00	-	N.A.	N.A.	N.A.	N.A.
28.00	-	N.A.	N.A.	N.A.	N.A.
30.00	-	N.A.	N.A.	N.A.	N.A.
32.00	-	N.A.	N.A.	N.A.	N.A.
34.00	-	N.A.	N.A.	N.A.	N.A.
36.00	-	N.A.	N.A.	N.A.	N.A.
38.00	-	N.A.	N.A.	N.A.	N.A.
40.00	-	N.A.	N.A.	N.A.	N.A.
42.00	-	N.A.	N.A.	N.A.	N.A.

NOTE (1) : FACTORS USED CD = 1.0, CF = VARIES BASED ON SIZE, CM = 1.0, Ct = 1.0, Ci = 1.0, CT = 1.0

NOTE (2): MEMBER OWN WEIGHT HAS BEEN FACTORED INTO THE CALCULATIONS

NOTE (3): #2 WOOD IS THE MOST COMMONLY AVAILABLE GRADE COMMERCIALLY

NOTE (4): N.A. INDICATES THAT SIZE OF TIMBER IS NOT COMMONLY AVAILABLE

TABLE 7-32

SAWN CUT TIMBER POSTS STRUCTURAL CAPACITY					
#2 EASTERN HEMLOCK-TAMARACK					
COLUMN CLEAR HEIGHT (FT)	4x4 POST STRUCTURAL CAPACITY (KIPS)(1),(2),(3)	6x6 POST STRUCTURAL CAPACITY (KIPS)(1),(2),(3)	8x8 POST STRUCTURAL CAPACITY (KIPS)(1),(2),(3)	10x10 POST STRUCTURAL CAPACITY (KIPS)(1),(2),(3)	12x12 POST STRUCTURAL CAPACITY (KIPS)(1),(2),(3)
4.00	9.92	11.79	22.21	35.82	52.63
6.00	7.20	11.37	21.82	35.46	52.23
8.00	4.72	10.67	21.26	34.92	51.70
10.00	3.19	9.64	20.36	34.11	51.04
12.00	2.27	8.28	19.18	33.12	50.12
14.00	1.70	6.92	17.60	31.76	48.93
16.00	-	5.68	15.80	30.05	47.34
18.00	-	4.68	13.89	27.97	45.49
20.00	-	3.93	12.03	25.63	43.24
22.00	-	3.29	10.46	23.19	40.73
24.00	-	-	9.05	20.84	37.82
26.00	-	-	7.93	18.59	34.91
28.00	-	-	6.97	16.60	32.00
30.00	-	-	6.13	14.89	29.22
32.00	-	-	-	13.35	26.58
34.00	-	-	-	12.00	24.33
36.00	-	-	-	10.83	22.08
38.00	-	-	-	9.83	20.23
40.00	-	-	-	-	18.51
42.00	-	-	-	-	17.06

NOTE (1) : FACTORS USED CD = 1.0, CF = VARIES BASED ON SIZE, CM = 1.0, Ct = 1.0, Ci = 1.0, CT = 1.0

NOTE (2): MEMBER OWN WEIGHT HAS BEEN FACTORED INTO THE CALCULATIONS

NOTE (3): #2 WOOD IS THE MOST COMMONLY AVAILABLE GRADE COMMERCIALLY

NOTE (4): N.A. INDICATES THAT SIZE OF TIMBER IS NOT COMMONLY AVAILABLE

SAWN CUT TIMBER POSTS STRUCTURAL CAPACITY					
#2 EASTERN SOFTWOODS					
COLUMN CLEAR HEIGHT (FT)	4x4 POST STRUCTURAL CAPACITY (KIPS)(1),(2),(3)	6x6 POST STRUCTURAL CAPACITY (KIPS)(1),(2),(3)	8x8 POST STRUCTURAL CAPACITY (KIPS)(1),(2),(3)	10x10 POST STRUCTURAL CAPACITY (KIPS)(1),(2),(3)	12x12 POST STRUCTURAL CAPACITY (KIPS)(1),(2),(3)
4.00	9.92	N.A.	N.A.	N.A.	N.A.
6.00	7.20	N.A.	N.A.	N.A.	N.A.
8.00	4.72	N.A.	N.A.	N.A.	N.A.
10.00	3.19	N.A.	N.A.	N.A.	N.A.
12.00	2.27	N.A.	N.A.	N.A.	N.A.
14.00	1.70	N.A.	N.A.	N.A.	N.A.
16.00	-	N.A.	N.A.	N.A.	N.A.
18.00	-	N.A.	N.A.	N.A.	N.A.
20.00	-	N.A.	N.A.	N.A.	N.A.
22.00	-	N.A.	N.A.	N.A.	N.A.
24.00	-	N.A.	N.A.	N.A.	N.A.
26.00	-	N.A.	N.A.	N.A.	N.A.
28.00	-	N.A.	N.A.	N.A.	N.A.
30.00	-	N.A.	N.A.	N.A.	N.A.
32.00	-	N.A.	N.A.	N.A.	N.A.
34.00	-	N.A.	N.A.	N.A.	N.A.
36.00	-	N.A.	N.A.	N.A.	N.A.
38.00	-	N.A.	N.A.	N.A.	N.A.
40.00	-	N.A.	N.A.	N.A.	N.A.
42.00	-	N.A.	N.A.	N.A.	N.A.

NOTE (1) : FACTORS USED CD = 1.0, CF = VARIES BASED ON SIZE, CM = 1.0, Ct = 1.0, Ci = 1.0, CT = 1.0

NOTE (2): MEMBER OWN WEIGHT HAS BEEN FACTORED INTO THE CALCULATIONS

NOTE (3): #2 WOOD IS THE MOST COMMONLY AVAILABLE GRADE COMMERCIALLY

NOTE (4): N.A. INDICATES THAT SIZE OF TIMBER IS NOT COMMONLY AVAILABLE

TABLE 7-34

SAWN CUT TIMBER POSTS STRUCTURAL CAPACITY					
#2 EASTERN WHITE PINE					
COLUMN CLEAR HEIGHT (FT)	4x4 POST STRUCTURAL CAPACITY (KIPS)(1),(2),(3)	6x6 POST STRUCTURAL CAPACITY (KIPS)(1),(2),(3)	8x8 POST STRUCTURAL CAPACITY (KIPS)(1),(2),(3)	10x10 POST STRUCTURAL CAPACITY (KIPS)(1),(2),(3)	12x12 POST STRUCTURAL CAPACITY (KIPS)(1),(2),(3)
4.00	9.92	9.64	18.11	29.15	42.84
6.00	7.20	9.37	17.83	28.88	42.58
8.00	4.72	8.92	17.49	28.51	42.18
10.00	3.19	8.28	16.93	28.06	41.79
12.00	2.27	7.41	16.20	27.43	41.12
14.00	1.70	6.38	15.18	26.62	40.46
16.00	-	5.38	14.00	25.54	39.41
18.00	-	4.53	12.65	24.18	38.35
20.00	-	3.81	11.25	22.65	36.89
22.00	-	3.23	9.90	21.02	35.31
24.00	-	-	8.71	19.22	33.45
26.00	-	-	7.65	17.50	31.47
28.00	-	-	6.75	15.79	29.35
30.00	-	-	6.01	14.25	27.11
32.00	-	-	-	12.90	24.99
34.00	-	-	-	11.64	23.01
36.00	-	-	-	10.55	21.16
38.00	-	-	-	9.65	19.44
40.00	-	-	-	-	17.98
42.00	-	-	-	-	16.53

NOTE (1) : FACTORS USED CD = 1.0, CF = VARIES BASED ON SIZE, CM = 1.0, Ct = 1.0, Ci = 1.0, CT = 1.0

NOTE (2): MEMBER OWN WEIGHT HAS BEEN FACTORED INTO THE CALCULATIONS

NOTE (3): #2 WOOD IS THE MOST COMMONLY AVAILABLE GRADE COMMERCIALLY

NOTE (4): N.A. INDICATES THAT SIZE OF TIMBER IS NOT COMMONLY AVAILABLE

TABLE 7-35

SAWN CUT TIMBER POSTS STRUCTURAL CAPACITY					
#2 HEM-FIR					
COLUMN CLEAR HEIGHT (FT)	4x4 POST STRUCTURAL CAPACITY (KIPS)(1),(2),(3)	6x6 POST STRUCTURAL CAPACITY (KIPS)(1),(2),(3)	8x8 POST STRUCTURAL CAPACITY (KIPS)(1),(2),(3)	10x10 POST STRUCTURAL CAPACITY (KIPS)(1),(2),(3)	12x12 POST STRUCTURAL CAPACITY (KIPS)(1),(2),(3)
4.00	14.47	16.87	31.83	51.44	75.51
6.00	9.34	16.15	31.21	50.81	74.98
8.00	5.81	14.91	30.15	49.81	74.06
10.00	3.85	13.06	28.63	48.46	72.86
12.00	2.73	10.92	26.49	46.65	71.15
14.00	2.02	8.86	23.79	44.22	69.03
16.00	-	7.19	20.81	41.24	66.25
18.00	-	5.86	17.94	37.72	62.95
20.00	-	4.87	15.35	33.93	59.11
22.00	-	4.08	13.16	30.23	54.75
24.00	-	-	11.36	26.71	50.12
26.00	-	-	9.84	23.64	45.62
28.00	-	-	8.60	20.93	41.39
30.00	-	-	7.59	18.68	37.29
32.00	-	-	-	16.69	33.72
34.00	-	-	-	14.98	30.54
36.00	-	-	-	13.44	27.77
38.00	-	-	-	12.18	25.25
40.00	-	-	-	-	23.01
42.00	-	-	-	-	21.16

NOTE (1) : FACTORS USED CD = 1.0, CF = VARIES BASED ON SIZE, CM = 1.0, Ct = 1.0, Ci = 1.0, CT = 1.0

NOTE (2): MEMBER OWN WEIGHT HAS BEEN FACTORED INTO THE CALCULATIONS

NOTE (3): #2 WOOD IS THE MOST COMMONLY AVAILABLE GRADE COMMERCIALLY

NOTE (4): N.A. INDICATES THAT SIZE OF TIMBER IS NOT COMMONLY AVAILABLE

TABLE 7-36

SAWN CUT TIMBER POSTS STRUCTURAL CAPACITY					
#2 HEM-FIR (NORTH)					
COLUMN CLEAR HEIGHT (FT)	4x4 POST STRUCTURAL CAPACITY (KIPS)(1),(2),(3)	6x6 POST STRUCTURAL CAPACITY (KIPS)(1),(2),(3)	8x8 POST STRUCTURAL CAPACITY (KIPS)(1),(2),(3)	10x10 POST STRUCTURAL CAPACITY (KIPS)(1),(2),(3)	12x12 POST STRUCTURAL CAPACITY (KIPS)(1),(2),(3)
4.00	16.64	16.87	31.83	51.44	75.51
6.00	11.19	16.15	31.21	50.81	74.98
8.00	7.06	14.91	30.15	49.81	74.06
10.00	4.72	13.06	28.63	48.46	72.86
12.00	3.34	10.92	26.49	46.65	71.15
14.00	2.48	8.86	23.79	44.22	69.03
16.00	-	7.19	20.81	41.24	66.25
18.00	-	5.86	17.94	37.72	62.95
20.00	-	4.87	15.35	33.93	59.11
22.00	-	4.08	13.16	30.23	54.75
24.00	-	-	11.36	26.71	50.12
26.00	-	-	9.84	23.64	45.62
28.00	-	-	8.60	20.93	41.39
30.00	-	-	7.59	18.68	37.29
32.00	-	-	-	16.69	33.72
34.00	-	-	-	14.98	30.54
36.00	-	-	-	13.44	27.77
38.00	-	-	-	12.18	25.25
40.00	-	-	-	-	23.01
42.00	-	-	-	-	21.16

NOTE (1) : FACTORS USED CD = 1.0, CF = VARIES BASED ON SIZE, CM = 1.0, Ct = 1.0, Ci = 1.0, CT = 1.0

NOTE (2): MEMBER OWN WEIGHT HAS BEEN FACTORED INTO THE CALCULATIONS

NOTE (3): #2 WOOD IS THE MOST COMMONLY AVAILABLE GRADE COMMERCIALLY

NOTE (4): N.A. INDICATES THAT SIZE OF TIMBER IS NOT COMMONLY AVAILABLE

TABLE 7-37

SAWN CUT TIMBER POSTS STRUCTURAL CAPACITY					
#2 MIXED MAPLE					
COLUMN CLEAR HEIGHT (FT)	4x4 POST STRUCTURAL CAPACITY (KIPS)(1),(2),(3)	6x6 POST STRUCTURAL CAPACITY (KIPS)(1),(2),(3)	8x8 POST STRUCTURAL CAPACITY (KIPS)(1),(2),(3)	10x10 POST STRUCTURAL CAPACITY (KIPS)(1),(2),(3)	12x12 POST STRUCTURAL CAPACITY (KIPS)(1),(2),(3)
4.00	7.05	8.92	16.70	26.89	39.54
6.00	5.88	8.68	16.48	26.71	39.27
8.00	4.29	8.31	16.20	26.44	39.01
10.00	3.05	7.77	15.75	25.99	38.61
12.00	2.21	7.04	15.13	25.45	38.08
14.00	1.66	6.17	14.28	24.72	37.55
16.00	-	5.26	13.27	23.91	36.76
18.00	-	4.44	12.15	22.83	35.70
20.00	-	3.75	10.91	21.47	34.64
22.00	-	3.20	9.67	20.03	33.32
24.00	-	-	8.55	18.50	31.74
26.00	-	-	7.53	16.96	30.02
28.00	-	-	6.69	15.43	28.16
30.00	-	-	5.96	13.98	26.18
32.00	-	-	-	12.72	24.33
34.00	-	-	-	11.55	22.48
36.00	-	-	-	10.46	20.76
38.00	-	-	-	9.56	19.17
40.00	-	-	-	-	17.72
42.00	-	-	-	-	16.27

NOTE (1) : FACTORS USED CD = 1.0, CF = VARIES BASED ON SIZE, CM = 1.0, Ct = 1.0, Ci = 1.0, CT = 1.0

NOTE (2): MEMBER OWN WEIGHT HAS BEEN FACTORED INTO THE CALCULATIONS

NOTE (3): #2 WOOD IS THE MOST COMMONLY AVAILABLE GRADE COMMERCIALLY

NOTE (4): N.A. INDICATES THAT SIZE OF TIMBER IS NOT COMMONLY AVAILABLE

SAWN CUT TIMBER POSTS STRUCTURAL CAPACITY					
#2 MIXED OAK					
COLUMN CLEAR HEIGHT (FT)	4x4 POST STRUCTURAL CAPACITY (KIPS)(1),(2),(3)	6x6 POST STRUCTURAL CAPACITY (KIPS)(1),(2),(3)	8x8 POST STRUCTURAL CAPACITY (KIPS)(1),(2),(3)	10x10 POST STRUCTURAL CAPACITY (KIPS)(1),(2),(3)	12x12 POST STRUCTURAL CAPACITY (KIPS)(1),(2),(3)
4.00	7.63	10.34	19.46	31.31	46.02
6.00	5.70	9.98	19.12	31.04	45.75
8.00	3.82	9.37	18.61	30.59	45.22
10.00	2.59	8.47	17.88	29.87	44.70
12.00	1.86	7.32	16.81	28.97	43.90
14.00	1.38	6.11	15.46	27.88	42.84
16.00	-	5.05	13.89	26.35	41.52
18.00	-	4.17	12.26	24.63	39.93
20.00	-	3.47	10.68	22.56	37.95
22.00	-	2.93	9.28	20.48	35.83
24.00	-	-	8.04	18.41	33.32
26.00	-	-	7.03	16.51	30.81
28.00	-	-	6.18	14.71	28.30
30.00	-	-	5.45	13.17	25.92
32.00	-	-	-	11.82	23.54
34.00	-	-	-	10.64	21.55
36.00	-	-	-	9.65	19.57
38.00	-	-	-	8.75	17.98
40.00	-	-	-	-	16.39
42.00	-	-	-	-	15.07

NOTE (1) : FACTORS USED CD = 1.0, CF = VARIES BASED ON SIZE, CM = 1.0, Ct = 1.0, Ci = 1.0, CT = 1.0

NOTE (2): MEMBER OWN WEIGHT HAS BEEN FACTORED INTO THE CALCULATIONS

NOTE (3): #2 WOOD IS THE MOST COMMONLY AVAILABLE GRADE COMMERCIALLY

NOTE (4): N.A. INDICATES THAT SIZE OF TIMBER IS NOT COMMONLY AVAILABLE

TABLE 7-39

SAWN CUT TIMBER POSTS STRUCTURAL CAPACITY					
#2 MIXED SOUTHERN PINE					
COLUMN CLEAR HEIGHT (FT)	4x4 POST STRUCTURAL CAPACITY (KIPS)[(1),(2),(3)]	6x6 POST STRUCTURAL CAPACITY (KIPS)[(1),(2),(3)]	8x8 POST STRUCTURAL CAPACITY (KIPS)[(1),(2),(3)]	10x10 POST STRUCTURAL CAPACITY (KIPS)[(1),(2),(3)]	12x12 POST STRUCTURAL CAPACITY (KIPS)[(1),(2),(3)]
4.00	14.50	15.42	29.08	46.93	69.03
6.00	9.77	14.73	28.46	46.38	68.50
8.00	6.18	13.61	27.50	45.48	67.57
10.00	4.12	11.91	26.10	44.22	66.52
12.00	2.92	9.92	24.13	42.59	64.93
14.00	2.18	8.07	21.65	40.34	62.95
16.00	-	6.53	18.95	37.54	60.43
18.00	-	5.35	16.31	34.29	57.39
20.00	-	4.41	14.00	30.86	53.82
22.00	-	3.72	11.98	27.52	49.85
24.00	-	-	10.35	24.36	45.75
26.00	-	-	9.00	21.56	41.52
28.00	-	-	78.75	19.13	37.55
30.00	-	-	6.91	16.96	33.98
32.00	-	-	-	15.16	30.68
34.00	-	-	-	13.62	27.77
36.00	-	-	-	12.27	25.25
38.00	-	-	-	11.10	23.01
40.00	-	-	-	-	21.02
42.00	-	-	-	-	19.17

NOTE (1) : FACTORS USED CD = 1.0, CF = VARIES BASED ON SIZE, CM = 1.0, Ct = 1.0, Ci = 1.0, CT = 1.0

NOTE (2): MEMBER OWN WEIGHT HAS BEEN FACTORED INTO THE CALCULATIONS

NOTE (3): #2 WOOD IS THE MOST COMMONLY AVAILABLE GRADE COMMERCIALLY

NOTE (4): N.A. INDICATES THAT SIZE OF TIMBER IS NOT COMMONLY AVAILABLE

TABLE 7-40

SAWN CUT TIMBER POSTS STRUCTURAL CAPACITY					
#2 NORTHERN RED OAK					
COLUMN CLEAR HEIGHT (FT)	4x4 POST STRUCTURAL CAPACITY (KIPS)(1),(2),(3)	6x6 POST STRUCTURAL CAPACITY (KIPS)(1),(2),(3)	8x8 POST STRUCTURAL CAPACITY (KIPS)(1),(2),(3)	10x10 POST STRUCTURAL CAPACITY (KIPS)(1),(2),(3)	12x12 POST STRUCTURAL CAPACITY (KIPS)(1),(2),(3)
4.00	9.17	11.82	22.21	35.82	52.63
6.00	7.39	11.46	21.88	35.55	52.37
8.00	5.24	10.85	21.37	35.01	51.84
10.00	3.66	9.95	20.64	34.38	51.18
12.00	2.64	8.74	19.57	33.48	50.38
14.00	1.98	7.41	18.22	32.30	49.32
16.00	-	6.17	16.53	30.77	48.00
18.00	-	5.14	14.73	28.97	46.41
20.00	-	4.29	12.99	26.89	44.43
22.00	-	3.63	11.30	24.54	42.18
24.00	-	-	9.90	22.29	39.67
26.00	-	-	8.66	20.12	36.89
28.00	-	-	7.65	18.05	34.12
30.00	-	-	6.75	16.24	31.34
32.00	-	-	-	14.62	28.69
34.00	-	-	-	13.17	26.31
36.00	-	-	-	11.91	24.06
38.00	-	-	-	10.83	22.08
40.00	-	-	-	-	20.23
42.00	-	-	-	-	18.64

NOTE (1) : FACTORS USED CD = 1.0, CF = VARIES BASED ON SIZE, CM = 1.0, Ct = 1.0, Ci = 1.0, CT = 1.0

NOTE (2): MEMBER OWN WEIGHT HAS BEEN FACTORED INTO THE CALCULATIONS

NOTE (3): #2 WOOD IS THE MOST COMMONLY AVAILABLE GRADE COMMERCIALLY

NOTE (4): N.A. INDICATES THAT SIZE OF TIMBER IS NOT COMMONLY AVAILABLE

TABLE 7-41

SAWN CUT TIMBER POSTS STRUCTURAL CAPACITY					
#2 NORTHERN SPECIES					
COLUMN CLEAR HEIGHT (FT)	4x4 POST STRUCTURAL CAPACITY (KIPS)(1),(2),(3)	6x6 POST STRUCTURAL CAPACITY (KIPS)(1),(2),(3)	8x8 POST STRUCTURAL CAPACITY (KIPS)(1),(2),(3)	10x10 POST STRUCTURAL CAPACITY (KIPS)(1),(2),(3)	12x12 POST STRUCTURAL CAPACITY (KIPS)(1),(2),(3)
4.00	10.15	N.A.	N.A.	N.A.	N.A.
6.00	7.28	N.A.	N.A.	N.A.	N.A.
8.00	4.75	N.A.	N.A.	N.A.	N.A.
10.00	3.20	N.A.	N.A.	N.A.	N.A.
12.00	2.27	N.A.	N.A.	N.A.	N.A.
14.00	1.70	N.A.	N.A.	N.A.	N.A.
16.00	-	N.A.	N.A.	N.A.	N.A.
18.00	-	N.A.	N.A.	N.A.	N.A.
20.00	-	N.A.	N.A.	N.A.	N.A.
22.00	-	N.A.	N.A.	N.A.	N.A.
24.00	-	-	N.A.	N.A.	N.A.
26.00	-	-	N.A.	N.A.	N.A.
28.00	-	-	N.A.	N.A.	N.A.
30.00	-	-	N.A.	N.A.	N.A.
32.00	-	-	-	N.A.	N.A.
34.00	-	-	-	N.A.	N.A.
36.00	-	-	-	N.A.	N.A.
38.00	-	-	-	N.A.	N.A.
40.00	-	-	-	-	N.A.
42.00	-	-	-	-	N.A.

NOTE (1) : FACTORS USED CD = 1.0, CF = VARIES BASED ON SIZE, CM = 1.0, Ct = 1.0, Ci = 1.0, CT = 1.0

NOTE (2): MEMBER OWN WEIGHT HAS BEEN FACTORED INTO THE CALCULATIONS

NOTE (3): #2 WOOD IS THE MOST COMMONLY AVAILABLE GRADE COMMERCIALLY

NOTE (4): N.A. INDICATES THAT SIZE OF TIMBER IS NOT COMMONLY AVAILABLE

TABLE 7-42

SAWN CUT TIMBER POSTS STRUCTURAL CAPACITY					
#2 NORTHERN WHITE CEDAR					
COLUMN CLEAR HEIGHT (FT)	4x4 POST STRUCTURAL CAPACITY (KIPS)(1),(2),(3)	6x6 POST STRUCTURAL CAPACITY (KIPS)(1),(2),(3)	8x8 POST STRUCTURAL CAPACITY (KIPS)(1),(2),(3)	10x10 POST STRUCTURAL CAPACITY (KIPS)(1),(2),(3)	12x12 POST STRUCTURAL CAPACITY (KIPS)(1),(2),(3)
4.00	5.83	7.38	13.89	22.38	32.93
6.00	4.39	7.13	13.66	22.20	32.66
8.00	2.95	6.74	13.33	21.84	32.40
10.00	2.02	6.14	12.82	21.38	32.00
12.00	1.44	5.35	12.15	20.84	31.47
14.00	1.07	4.50	11.25	20.03	30.68
16.00	-	3.75	10.18	19.04	29.88
18.00	-	3.11	9.00	17.86	28.83
20.00	-	2.60	7.87	16.51	27.50
22.00	-	2.17	6.86	15.07	26.05
24.00	-	-	5.96	13.62	24.33
26.00	-	-	5.23	12.18	22.61
28.00	-	-	4.61	10.92	20.76
30.00	-	-	4.05	9.83	19.04
32.00	-	-	-	8.84	17.45
34.00	-	-	-	7.94	16.00
36.00	-	-	-	7.22	14.54
38.00	-	-	-	6.49	13.35
40.00	-	-	-	-	12.29
42.00	-	-	-	-	11.24

NOTE (1) : FACTORS USED CD = 1.0, CF = VARIES BASED ON SIZE, CM = 1.0, Ct = 1.0, Ci = 1.0, CT = 1.0

NOTE (2): MEMBER OWN WEIGHT HAS BEEN FACTORED INTO THE CALCULATIONS

NOTE (3): #2 WOOD IS THE MOST COMMONLY AVAILABLE GRADE COMMERCIALLY

NOTE (4): N.A. INDICATES THAT SIZE OF TIMBER IS NOT COMMONLY AVAILABLE

TABLE 7-43

SAWN CUT TIMBER POSTS STRUCTURAL CAPACITY					
#2 RED MAPLE					
COLUMN CLEAR HEIGHT (FT)	4x4 POST STRUCTURAL CAPACITY (KIPS)(1),(2),(3)	6x6 POST STRUCTURAL CAPACITY (KIPS)(1),(2),(3)	8x8 POST STRUCTURAL CAPACITY (KIPS)(1),(2),(3)	10x10 POST STRUCTURAL CAPACITY (KIPS)(1),(2),(3)	12x12 POST STRUCTURAL CAPACITY (KIPS)(1),(2),(3)
4.00	9.05	11.16	20.92	33.66	49.46
6.00	7.66	10.89	20.64	33.39	49.19
8.00	5.73	10.46	20.30	33.12	48.80
10.00	4.10	9.86	19.80	32.58	48.40
12.00	3.00	8.98	19.06	31.94	47.87
14.00	2.26	7.95	18.11	31.13	47.08
16.00	-	6.86	16.98	30.14	46.15
18.00	-	5.83	15.58	28.88	45.09
20.00	-	4.96	14.11	27.43	43.77
22.00	-	4.23	12.60	25.72	42.18
24.00	-	-	11.19	23.91	40.46
26.00	-	-	9.95	22.02	38.35
28.00	-	-	8.83	20.12	36.23
30.00	-	-	7.87	18.32	33.98
32.00	-	-	-	16.69	31.60
34.00	-	-	-	15.16	29.35
36.00	-	-	-	13.80	27.24
38.00	-	-	-	12.63	25.12
40.00	-	-	-	-	23.27
42.00	-	-	-	-	21.55

NOTE (1) : FACTORS USED CD = 1.0, CF = VARIES BASED ON SIZE, CM = 1.0, Ct = 1.0, Ci = 1.0, CT = 1.0

NOTE (2): MEMBER OWN WEIGHT HAS BEEN FACTORED INTO THE CALCULATIONS

NOTE (3): #2 WOOD IS THE MOST COMMONLY AVAILABLE GRADE COMMERCIALLY

NOTE (4): N.A. INDICATES THAT SIZE OF TIMBER IS NOT COMMONLY AVAILABLE

TABLE 7-44

SAWN CUT TIMBER POSTS STRUCTURAL CAPACITY					
#2 RED OAK					
COLUMN CLEAR HEIGHT (FT)	4x4 POST STRUCTURAL CAPACITY (KIPS)(1),(2),(3)	6x6 POST STRUCTURAL CAPACITY (KIPS)(1),(2),(3)	8x8 POST STRUCTURAL CAPACITY (KIPS)(1),(2),(3)	10x10 POST STRUCTURAL CAPACITY (KIPS)(1),(2),(3)	12x12 POST STRUCTURAL CAPACITY (KIPS)(1),(2),(3)
4.00	7.98	10.37	19.51	31.40	46.15
6.00	6.56	10.10	19.23	31.13	45.89
8.00	4.75	9.64	18.84	30.77	45.49
10.00	3.34	8.98	18.28	30.23	44.96
12.00	2.42	8.07	17.49	29.60	44.43
14.00	1.82	6.98	16.53	28.69	43.64
16.00	-	5.92	15.24	27.61	42.58
18.00	-	4.99	13.83	26.26	41.39
20.00	-	4.20	12.31	24.72	40.07
22.00	-	3.56	10.91	22.92	38.35
24.00	-	-	9.61	21.02	36.36
26.00	-	-	8.49	19.22	34.25
28.00	-	-	7.48	17.41	32.00
30.00	-	-	6.63	15.70	29.75
32.00	-	-	-	14.25	27.50
34.00	-	-	-	12.90	25.39
36.00	-	-	-	11.73	23.40
38.00	-	-	-	10.64	21.55
40.00	-	-	-	-	19.83
42.00	-	-	-	-	18.25

NOTE (1) : FACTORS USED CD = 1.0, CF = VARIES BASED ON SIZE, CM = 1.0, Ct = 1.0, Ci = 1.0, CT = 1.0

NOTE (2): MEMBER OWN WEIGHT HAS BEEN FACTORED INTO THE CALCULATIONS

NOTE (3): #2 WOOD IS THE MOST COMMONLY AVAILABLE GRADE COMMERCIALLY

NOTE (4): N.A. INDICATES THAT SIZE OF TIMBER IS NOT COMMONLY AVAILABLE

TABLE 7-45

SAWN CUT TIMBER POSTS STRUCTURAL CAPACITY					
#2 REDWOOD					
COLUMN CLEAR HEIGHT (FT)	4x4 POST STRUCTURAL CAPACITY (KIPS)(1),(2),(3)	6x6 POST STRUCTURAL CAPACITY (KIPS)(1),(2),(3)	8x8 POST STRUCTURAL CAPACITY (KIPS)(1),(2),(3)	10x10 POST STRUCTURAL CAPACITY (KIPS)(1),(2),(3)	12x12 POST STRUCTURAL CAPACITY (KIPS)(1),(2),(3)
4.00	11.29	25.95	49.44	80.05	117.83
6.00	8.02	23.89	47.70	78.42	116.38
8.00	5.20	20.44	44.77	75.90	113.99
10.00	3.50	16.21	40.44	72.10	110.56
12.00	2.49	12.46	34.93	66.96	106.06
14.00	1.86	9.64	29.13	60.37	99.98
16.00	-	7.62	24.01	53.06	92.44
18.00	-	6.11	19.85	45.84	83.84
20.00	-	5.02	16.53	39.34	74.85
22.00	-	4.17	13.95	33.75	66.25
24.00	-	-	11.86	29.15	58.45
26.00	-	-	10.23	25.36	51.44
28.00	-	-	8.88	22.20	45.49
30.00	-	-	7.76	19.49	40.46
32.00	-	-	-	17.32	36.10
34.00	-	-	-	15.43	32.26
36.00	-	-	-	13.80	29.09
38.00	-	-	-	12.45	26.31
40.00	-	-	-	-	23.93
42.00	-	-	-	-	21.82

NOTE (1) : FACTORS USED CD = 1.0, CF = VARIES BASED ON SIZE, CM = 1.0, Ct = 1.0, Ci = 1.0, CT = 1.0

NOTE (2): MEMBER OWN WEIGHT HAS BEEN FACTORED INTO THE CALCULATIONS

NOTE (3): #2 WOOD IS THE MOST COMMONLY AVAILABLE GRADE COMMERCIALLY

NOTE (4): N.A. INDICATES THAT SIZE OF TIMBER IS NOT COMMONLY AVAILABLE

TABLE 7-46

SAWN CUT TIMBER POSTS STRUCTURAL CAPACITY					
#2 SOUTHERN PINE					
COLUMN CLEAR HEIGHT (FT)	4x4 POST STRUCTURAL CAPACITY (KIPS)(1),(2),(3)	6x6 POST STRUCTURAL CAPACITY (KIPS)(1),(2),(3)	8x8 POST STRUCTURAL CAPACITY (KIPS)(1),(2),(3)	10x10 POST STRUCTURAL CAPACITY (KIPS)(1),(2),(3)	12x12 POST STRUCTURAL CAPACITY (KIPS)(1),(2),(3)
4.00	14.50	15.51	29.19	47.02	69.03
6.00	9.77	14.97	28.68	46.56	68.63
8.00	6.18	14.06	27.90	45.84	67.97
10.00	4.12	12.70	26.77	44.85	67.05
12.00	2.92	10.98	25.25	43.50	65.86
14.00	2.18	9.16	23.23	41.78	64.27
16.00	-	7.56	20.86	39.52	62.28
18.00	-	6.26	18.39	36.91	59.90
20.00	-	5.20	16.03	33.93	56.99
22.00	-	4.38	13.89	30.77	53.69
24.00	-	-	12.09	27.61	50.12
26.00	-	-	10.51	24.72	46.28
28.00	-	-	9.28	22.11	42.45
30.00	-	-	8.15	19.76	38.74
32.00	-	-	-	17.77	35.44
34.00	-	-	-	15.97	32.26
36.00	-	-	-	14.44	29.49
38.00	-	-	-	13.08	26.84
40.00	-	-	-	-	24.59
42.00	-	-	-	-	22.61

NOTE (1) : FACTORS USED CD = 1.0, CF = VARIES BASED ON SIZE, CM = 1.0, Ct = 1.0, Ci = 1.0, CT = 1.0

NOTE (2): MEMBER OWN WEIGHT HAS BEEN FACTORED INTO THE CALCULATIONS

NOTE (3): #2 WOOD IS THE MOST COMMONLY AVAILABLE GRADE COMMERCIALLY

NOTE (4): N.A. INDICATES THAT SIZE OF TIMBER IS NOT COMMONLY AVAILABLE

TABLE 7-47

SAWN CUT TIMBER POSTS STRUCTURAL CAPACITY					
#2 SPRUCE-PINE-FIR (SOUTH)					
COLUMN CLEAR HEIGHT (FT)	4x4 POST STRUCTURAL CAPACITY (KIPS)(1),(2),(3)	6x6 POST STRUCTURAL CAPACITY (KIPS)(1),(2),(3)	8x8 POST STRUCTURAL CAPACITY (KIPS)(1),(2),(3)	10x10 POST STRUCTURAL CAPACITY (KIPS)(1),(2),(3)	12x12 POST STRUCTURAL CAPACITY (KIPS)(1),(2),(3)
4.00	11.46	12.55	23.62	38.08	55.94
6.00	7.70	12.13	23.23	37.72	55.54
8.00	4.86	11.43	22.61	37.18	55.01
10.00	3.24	10.37	21.76	36.37	54.35
12.00	2.30	9.01	20.58	35.37	53.29
14.00	1.71	7.56	19.01	33.93	52.10
16.00	-	6.26	17.10	32.21	50.65
18.00	-	5.17	15.13	30.14	48.80
20.00	-	4.32	13.21	27.79	46.55
22.00	-	3.66	11.47	25.27	43.90
24.00	-	-	10.01	22.83	40.99
26.00	-	-	8.77	20.48	38.08
28.00	-	-	7.70	18.32	35.04
30.00	-	-	6.80	16.42	32.00
32.00	-	-	-	14.71	29.22
34.00	-	-	-	13.26	26.71
36.00	-	-	-	12.00	24.33
38.00	-	-	-	10.92	22.35
40.00	-	-	-	-	20.49
42.00	-	-	-	-	18.77

NOTE (1) : FACTORS USED CD = 1.0, CF = VARIES BASED ON SIZE, CM = 1.0, Ct = 1.0, Ci = 1.0, CT = 1.0

NOTE (2): MEMBER OWN WEIGHT HAS BEEN FACTORED INTO THE CALCULATIONS

NOTE (3): #2 WOOD IS THE MOST COMMONLY AVAILABLE GRADE COMMERCIALLY

NOTE (4): N.A. INDICATES THAT SIZE OF TIMBER IS NOT COMMONLY AVAILABLE

TABLE 7-48

SAWN CUT TIMBER POSTS STRUCTURAL CAPACITY					
#2 SPRUCE-PINE-FIR					
COLUMN CLEAR HEIGHT (FT)	4x4 POST STRUCTURAL CAPACITY (KIPS)(1),(2),(3)	6x6 POST STRUCTURAL CAPACITY (KIPS)(1),(2),(3)	8x8 POST STRUCTURAL CAPACITY (KIPS)(1),(2),(3)	10x10 POST STRUCTURAL CAPACITY (KIPS)(1),(2),(3)	12x12 POST STRUCTURAL CAPACITY (KIPS)(1),(2),(3)
4.00	13.55	14.70	27.73	44.76	65.73
6.00	9.48	14.10	27.17	44.22	65.20
8.00	6.10	13.07	26.33	43.41	64.54
10.00	4.10	11.56	25.03	42.33	63.48
12.00	2.92	9.74	23.29	40.79	62.16
14.00	2.17	7.96	21.09	38.81	60.31
16.00	-	6.47	18.56	36.37	58.06
18.00	-	5.32	16.09	33.39	55.41
20.00	-	4.42	13.84	30.23	52.11
22.00	-	3.69	11.87	26.98	48.54
24.00	-	-	10.29	24.01	44.70
26.00	-	-	8.94	21.30	40.87
28.00	-	-	7.82	18.95	37.03
30.00	-	-	6.86	16.88	33.59
32.00	-	-	-	15.07	30.42
34.00	-	-	-	13.54	27.64
36.00	-	-	-	12.18	25.13
38.00	-	-	-	11.01	22.88
40.00	-	-	-	-	20.90
42.00	-	-	-	-	19.18

NOTE (1) : FACTORS USED CD = 1.0, CF = VARIES BASED ON SIZE, CM = 1.0, Ct = 1.0, Ci = 1.0, CT = 1.0

NOTE (2): MEMBER OWN WEIGHT HAS BEEN FACTORED INTO THE CALCULATIONS

NOTE (3): #2 WOOD IS THE MOST COMMONLY AVAILABLE GRADE COMMERCIALLY

NOTE (4): N.A. INDICATES THAT SIZE OF TIMBER IS NOT COMMONLY AVAILABLE

TABLE 7-49

SAWN CUT TIMBER POSTS STRUCTURAL CAPACITY					
#2 WESTERN CEDARS					
COLUMN CLEAR HEIGHT (FT)	4x4 POST STRUCTURAL CAPACITY (KIPS)$^{(1),(2),(3)}$	6x6 POST STRUCTURAL CAPACITY (KIPS)$^{(1),(2),(3)}$	8x8 POST STRUCTURAL CAPACITY (KIPS)$^{(1),(2),(3)}$	10x10 POST STRUCTURAL CAPACITY (KIPS)$^{(1),(2),(3)}$	12x12 POST STRUCTURAL CAPACITY (KIPS)$^{(1),(2),(3)}$
4.00	8.04	16.00	30.32	49.01	72.21
6.00	6.16	15.00	29.48	48.19	71.42
8.00	4.19	13.28	28.07	46.93	70.22
10.00	2.87	10.98	25.93	45.13	68.51
12.00	2.06	8.68	23.06	42.60	66.26
14.00	1.53	6.84	19.86	39.35	63.35
16.00	-	5.45	16.71	35.38	59.64
18.00	-	4.39	14.01	31.32	55.28
20.00	-	3.63	11.76	27.26	50.39
22.00	-	3.03	9.96	23.74	45.36
24.00	-	-	8.49	20.67	40.47
26.00	-	-	7.37	18.05	36.10
28.00	-	-	6.41	15.88	32.14
30.00	-	-	5.63	13.99	28.70
32.00	-	-	-	12.45	25.66
34.00	-	-	-	11.10	23.14
36.00	-	-	-	10.02	20.90
38.00	-	-	-	9.03	18.91
40.00	-	-	-	-	17.19
42.00	-	-	-	-	15.74

NOTE (1) : FACTORS USED CD = 1.0, CF = VARIES BASED ON SIZE, CM = 1.0, Ct = 1.0, Ci = 1.0, CT = 1.0

NOTE (2): MEMBER OWN WEIGHT HAS BEEN FACTORED INTO THE CALCULATIONS

NOTE (3): #2 WOOD IS THE MOST COMMONLY AVAILABLE GRADE COMMERCIALLY

NOTE (4): N.A. INDICATES THAT SIZE OF TIMBER IS NOT COMMONLY AVAILABLE

TABLE 7-50

SAWN CUT TIMBER POSTS STRUCTURAL CAPACITY					
#2 WESTERN WOODS					
COLUMN CLEAR HEIGHT (FT)	4x4 POST STRUCTURAL CAPACITY (KIPS)(1),(2),(3)	6x6 POST STRUCTURAL CAPACITY (KIPS)(1),(2),(3)	8x8 POST STRUCTURAL CAPACITY (KIPS)(1),(2),(3)	10x10 POST STRUCTURAL CAPACITY (KIPS)(1),(2),(3)	12x12 POST STRUCTURAL CAPACITY (KIPS)(1),(2),(3)
4.00	10.35	13.95	26.33	42.51	62.42
6.00	6.98	13.34	25.76	41.97	61.89
8.00	4.42	12.28	24.92	41.15	61.10
10.00	2.95	10.77	23.63	40.07	60.17
12.00	2.09	8.95	21.83	38.45	58.72
14.00	1.56	7.26	19.58	36.46	57.00
16.00	-	5.90	17.10	33.93	54.75
18.00	-	4.81	14.74	30.96	51.97
20.00	-	3.99	12.60	27.89	48.67
22.00	-	3.36	10.80	24.82	44.97
24.00	-	-	9.34	21.93	41.26
26.00	-	-	8.10	19.40	37.43
28.00	-	-	7.09	17.24	33.86
30.00	-	-	6.24	15.25	30.68
32.00	-	-	-	13.63	27.64
34.00	-	-	-	12.27	25.13
36.00	-	-	-	11.01	22.75
38.00	-	-	-	10.02	20.76
40.00	-	-	-	-	18.91
42.00	-	-	-	-	17.32

NOTE (1) : FACTORS USED CD = 1.0, CF = VARIES BASED ON SIZE, CM = 1.0, Ct = 1.0, Ci = 1.0, CT = 1.0

NOTE (2): MEMBER OWN WEIGHT HAS BEEN FACTORED INTO THE CALCULATIONS

NOTE (3): #2 WOOD IS THE MOST COMMONLY AVAILABLE GRADE COMMERCIALLY

NOTE (4): N.A. INDICATES THAT SIZE OF TIMBER IS NOT COMMONLY AVAILABLE

TABLE 7-51

SAWN CUT TIMBER POSTS STRUCTURAL CAPACITY					
#2 WHITE OAK					
COLUMN CLEAR HEIGHT (FT)	4x4 POST STRUCTURAL CAPACITY (KIPS)(1),(2),(3)	6x6 POST STRUCTURAL CAPACITY (KIPS)(1),(2),(3)	8x8 POST STRUCTURAL CAPACITY (KIPS)(1),(2),(3)	10x10 POST STRUCTURAL CAPACITY (KIPS)(1),(2),(3)	12x12 POST STRUCTURAL CAPACITY (KIPS)(1),(2),(3)
4.00	8.35	11.77	22.16	35.83	52.64
6.00	5.98	11.28	21.77	35.38	52.24
8.00	3.90	10.47	21.04	34.75	51.58
10.00	2.62	9.26	20.03	33.84	50.78
12.00	1.87	7.77	18.62	32.67	49.73
14.00	1.36	6.38	16.88	31.05	48.27
16.00	-	5.17	14.85	29.06	46.55
18.00	-	4.27	12.88	26.71	44.30
20.00	-	3.54	11.08	24.19	41.79
22.00	-	2.96	9.51	21.66	38.88
24.00	-	-	8.21	19.22	35.71
26.00	-	-	7.14	17.06	32.67
28.00	-	-	6.24	15.16	29.62
30.00	-	-	5.51	13.45	26.85
32.00	-	-	-	12.09	24.33
34.00	-	-	-	10.83	22.09
36.00	-	-	-	9.75	20.10
38.00	-	-	-	8.84	18.25
40.00	-	-	-	-	16.66
42.00	-	-	-	-	15.34

NOTE (1) : FACTORS USED CD = 1.0, CF = VARIES BASED ON SIZE, CM = 1.0, Ct = 1.0, Ci = 1.0, CT = 1.0

NOTE (2): MEMBER OWN WEIGHT HAS BEEN FACTORED INTO THE CALCULATIONS

NOTE (3): #2 WOOD IS THE MOST COMMONLY AVAILABLE GRADE COMMERCIALLY

NOTE (4): N.A. INDICATES THAT SIZE OF TIMBER IS NOT COMMONLY AVAILABLE

TABLE 7-52

SAWN CUT TIMBER POSTS STRUCTURAL CAPACITY					
#2 YELLOW CEDAR					
COLUMN CLEAR HEIGHT (FT)	4x4 POST STRUCTURAL CAPACITY (KIPS)(1),(2),(3)	6x6 POST STRUCTURAL CAPACITY (KIPS)(1),(2),(3)	8x8 POST STRUCTURAL CAPACITY (KIPS)(1),(2),(3)	10x10 POST STRUCTURAL CAPACITY (KIPS)(1),(2),(3)	12x12 POST STRUCTURAL CAPACITY (KIPS)(1),(2),(3)
4.00	12.15	N.A.	N.A.	N.A.	N.A.
6.00	8.99	N.A.	N.A.	N.A.	N.A.
8.00	5.97	N.A.	N.A.	N.A.	N.A.
10.00	4.05	N.A.	N.A.	N.A.	N.A.
12.00	2.89	N.A.	N.A.	N.A.	N.A.
14.00	2.16	N.A.	N.A.	N.A.	N.A.
16.00	-	N.A.	N.A.	N.A.	N.A.
18.00	-	N.A.	N.A.	N.A.	N.A.
20.00	-	N.A.	N.A.	N.A.	N.A.
22.00	-	N.A.	N.A.	N.A.	N.A.
24.00	-	-	N.A.	N.A.	N.A.
26.00	-	-	N.A.	N.A.	N.A.
28.00	-	-	N.A.	N.A.	N.A.
30.00	-	-	N.A.	N.A.	N.A.
32.00	-	-	-	N.A.	N.A.
34.00	-	-	-	N.A.	N.A.
36.00	-	-	-	N.A.	N.A.
38.00	-	-	-	N.A.	N.A.
40.00	-	-	-	-	N.A.
42.00	-	-	-	-	N.A.

NOTE (1) : FACTORS USED CD = 1.0, CF = VARIES BASED ON SIZE, CM = 1.0, Ct = 1.0, Ci = 1.0, CT = 1.0

NOTE (2): MEMBER OWN WEIGHT HAS BEEN FACTORED INTO THE CALCULATIONS

NOTE (3): #2 WOOD IS THE MOST COMMONLY AVAILABLE GRADE COMMERCIALLY

NOTE (4): N.A. INDICATES THAT SIZE OF TIMBER IS NOT COMMONLY AVAILABLE

SAWN CUT TIMBER POSTS STRUCTURAL CAPACITY					
#2 YELLOW POPLAR					
COLUMN CLEAR HEIGHT (FT)	4x4 POST STRUCTURAL CAPACITY (KIPS)(1),(2),(3)	6x6 POST STRUCTURAL CAPACITY (KIPS)(1),(2),(3)	8x8 POST STRUCTURAL CAPACITY (KIPS)(1),(2),(3)	10x10 POST STRUCTURAL CAPACITY (KIPS)(1),(2),(3)	12x12 POST STRUCTURAL CAPACITY (KIPS)(1),(2),(3)
4.00	7.47	N.A.	N.A.	N.A.	N.A.
6.00	6.42	N.A.	N.A.	N.A.	N.A.
8.00	4.88	N.A.	N.A.	N.A.	N.A.
10.00	3.53	N.A.	N.A.	N.A.	N.A.
12.00	2.58	N.A.	N.A.	N.A.	N.A.
14.00	1.98	N.A.	N.A.	N.A.	N.A.
16.00	-	N.A.	N.A.	N.A.	N.A.
18.00	-	N.A.	N.A.	N.A.	N.A.
20.00	-	N.A.	N.A.	N.A.	N.A.
22.00	-	N.A.	N.A.	N.A.	N.A.
24.00	-	-	N.A.	N.A.	N.A.
26.00	-	-	N.A.	N.A.	N.A.
28.00	-	-	N.A.	N.A.	N.A.
30.00	-	-	N.A.	N.A.	N.A.
32.00	-	-	-	N.A.	N.A.
34.00	-	-	-	N.A.	N.A.
36.00	-	-	-	N.A.	N.A.
38.00	-	-	-	N.A.	N.A.
40.00	-	-	-	-	N.A.
42.00	-	-	-	-	N.A.

NOTE (1) : FACTORS USED CD = 1.0, CF = VARIES BASED ON SIZE, CM = 1.0, Ct = 1.0, Ci = 1.0, CT = 1.0

NOTE (2): MEMBER OWN WEIGHT HAS BEEN FACTORED INTO THE CALCULATIONS

NOTE (3): #2 WOOD IS THE MOST COMMONLY AVAILABLE GRADE COMMERCIALLY

NOTE (4): N.A. INDICATES THAT SIZE OF TIMBER IS NOT COMMONLY AVAILABLE

WELDS

BASIC WELDING SYMBOLS AND THEIR LOCATION SIGNIFICANCE

LOCATION SIGNIFICANCE	FILLET	PLUG OR SLOT
ARROW SIDE	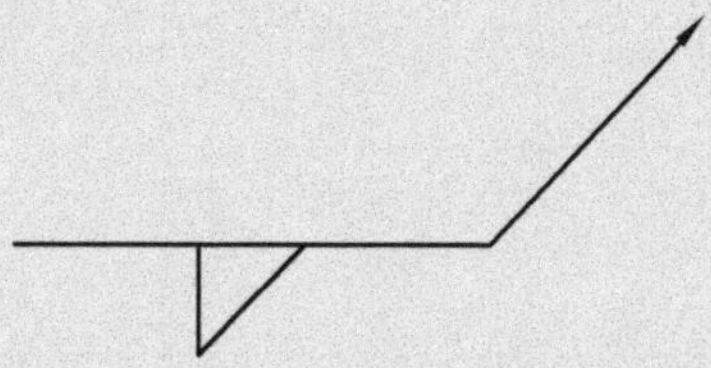	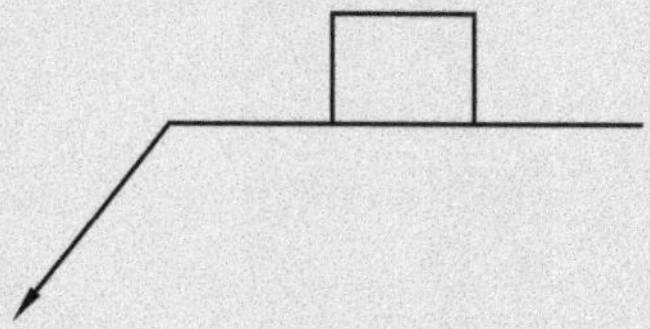
OTHER SIDE	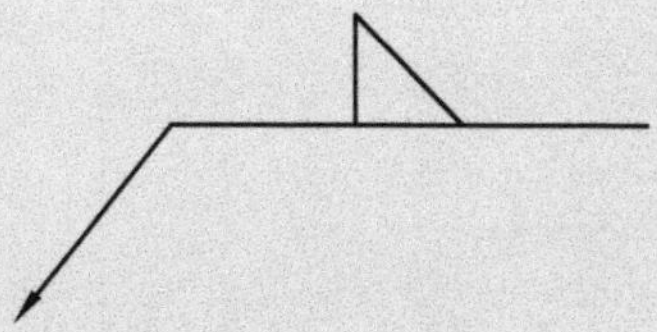	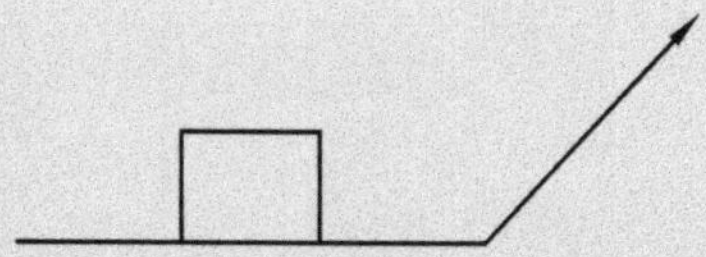
BOTH SIDES	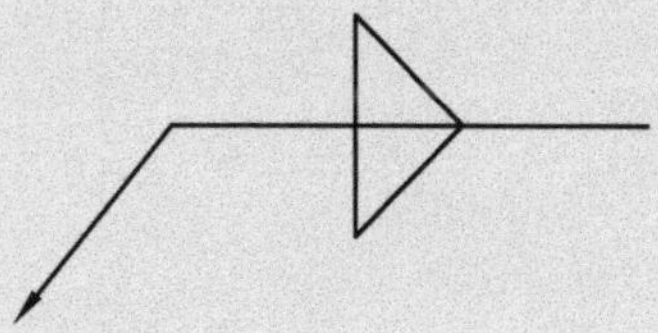	NOT USED
NO ARROW SIDE OR OTHER SIDE SIGNIFICANCE	NOT USED	NOT USED

REFERENCE 8-2

BASIC WELDING SYMBOLS AND THEIR LOCATION SIGNIFICANCE

LOCATION SIGNIFICANCE	SPOT OR PROJECTION	STUD
ARROW SIDE	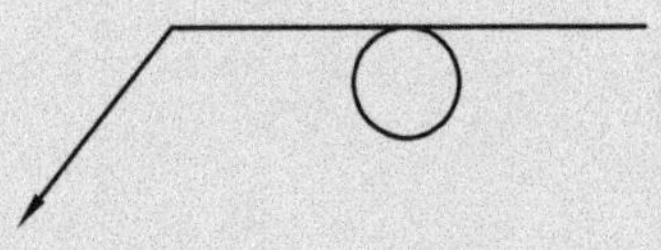	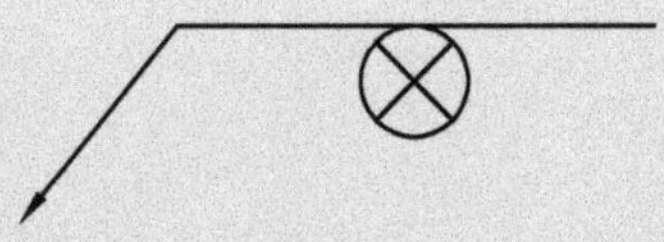
OTHER SIDE	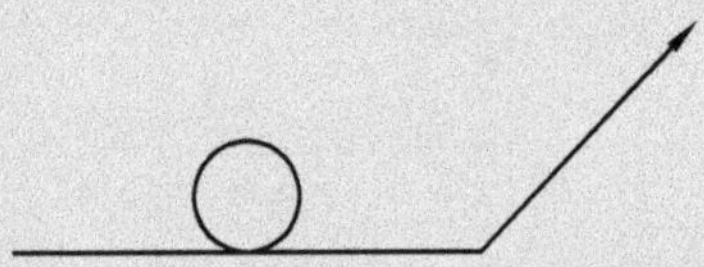	NOT USED
BOTH SIDES	NOT USED	NOT USED
NO ARROW SIDE OR OTHER SIDE SIGNIFICANCE	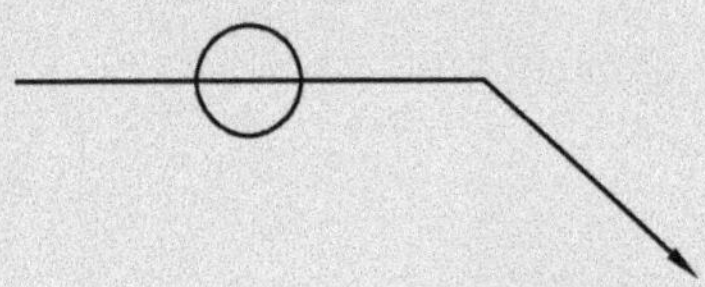	NOT USED

BASIC WELDING SYMBOLS AND THEIR LOCATION SIGNIFICANCE

LOCATION SIGNIFICANCE	SEAM	BACK OR BACKING
ARROW SIDE	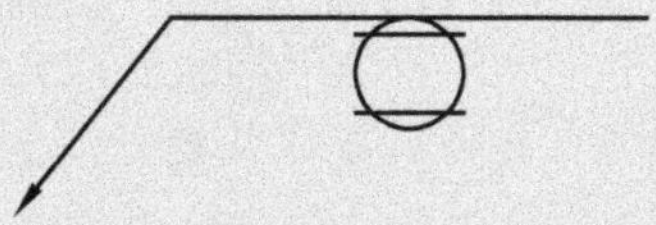	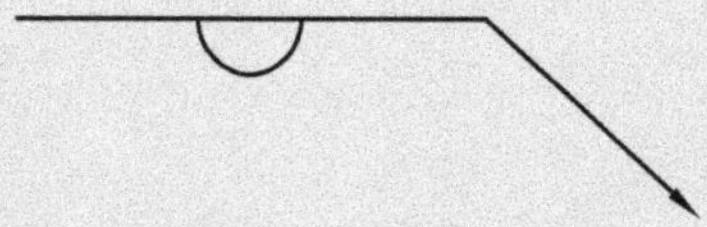
OTHER SIDE	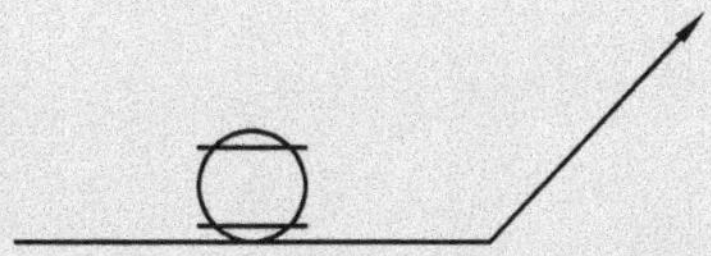	
BOTH SIDES	NOT USED	NOT USED
NO ARROW SIDE OR OTHER SIDE SIGNIFICANCE	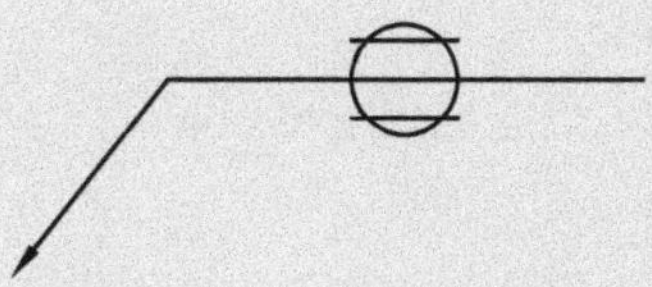	NOT USED

BASIC WELDING SYMBOLS AND THEIR LOCATION SIGNIFICANCE

LOCATION SIGNIFICANCE	SURFACING	EDGE
ARROW SIDE	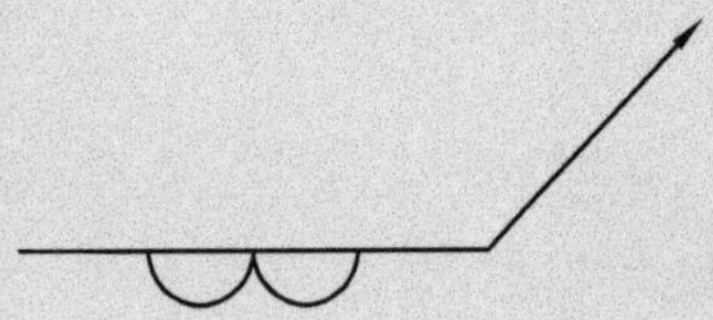	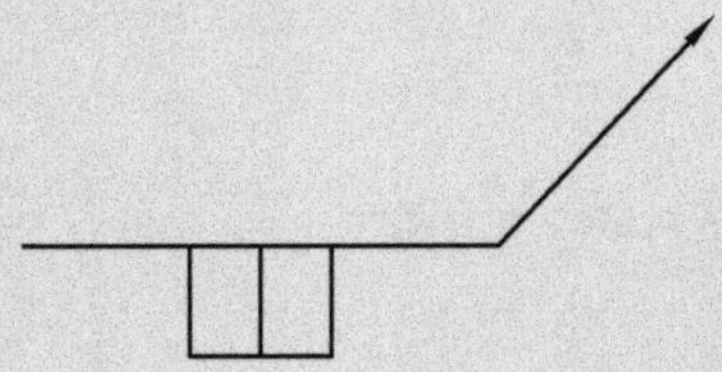
OTHER SIDE	NOT USED	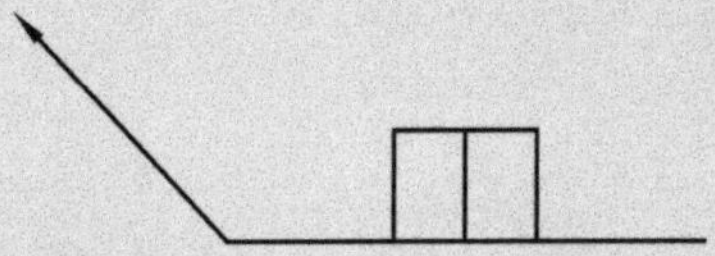
BOTH SIDES	NOT USED	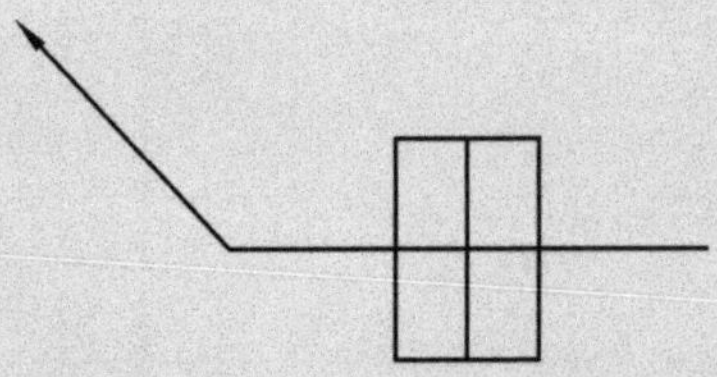
NO ARROW SIDE OR OTHER SIDE SIGNIFICANCE	NOT USED	NOT USED

REFERENCE 8-5

GROVE WELDING SYMBOLS AND THEIR LOCATION SIGNIFICANCE

LOCATION SIGNIFICANCE	SQUARE	V
ARROW SIDE	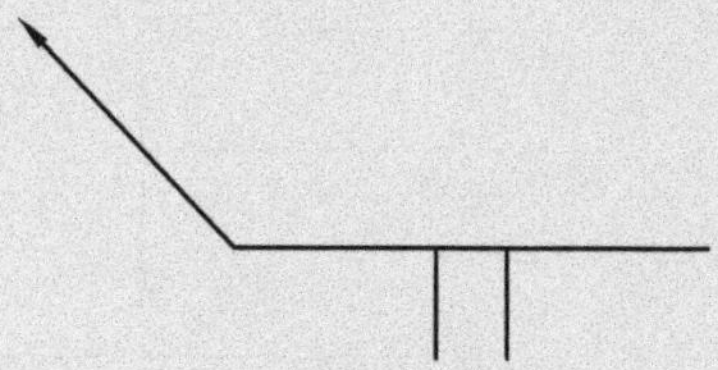	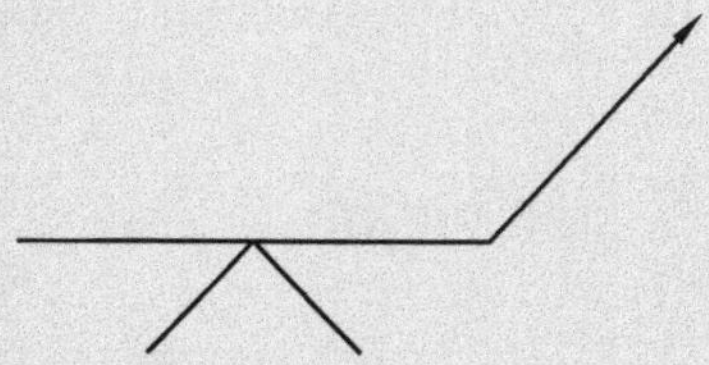
OTHER SIDE	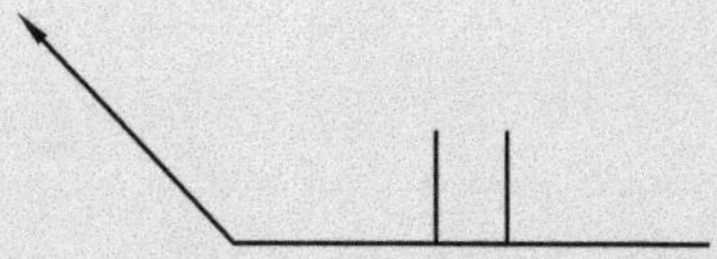	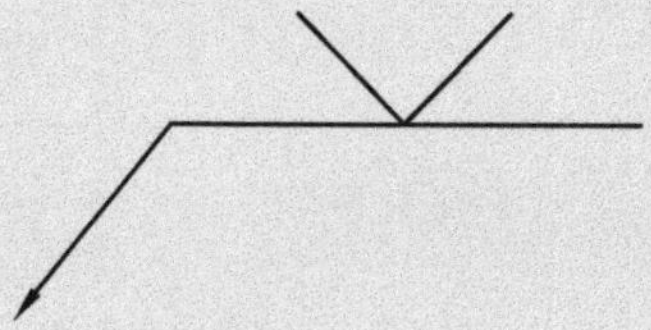
BOTH SIDES	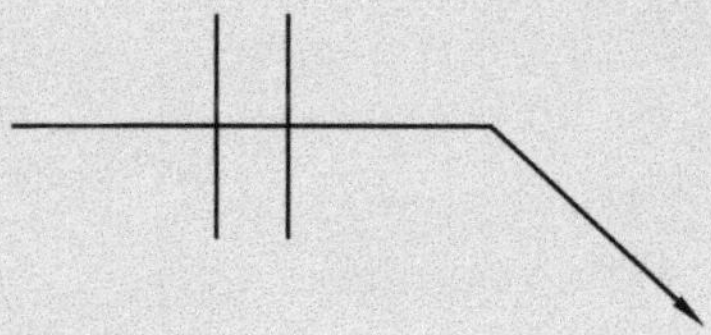	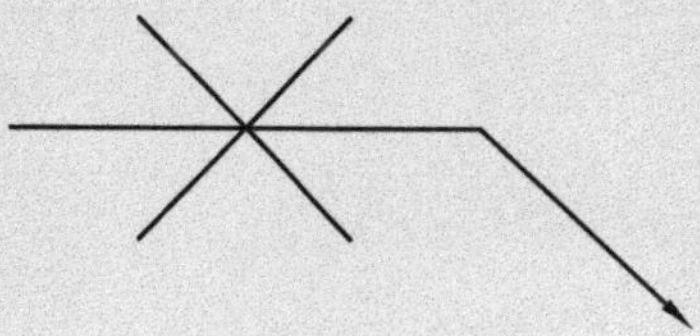
NO ARROW SIDE OR OTHER SIDE SIGNIFICANCE	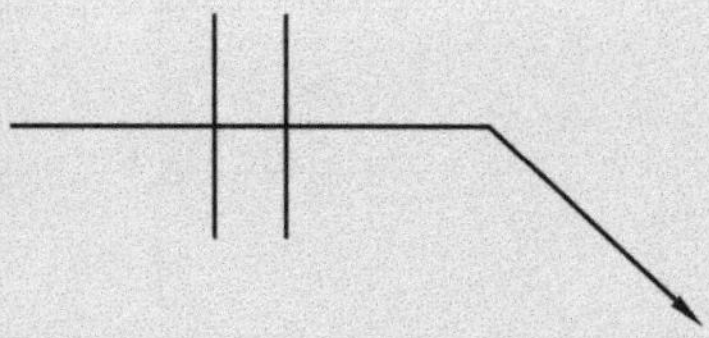	NOT USED

REFERENCE 8-6

GROVE WELDING SYMBOLS AND THEIR LOCATION SIGNIFICANCE

LOCATION SIGNIFICANCE	BEVEL	U
ARROW SIDE	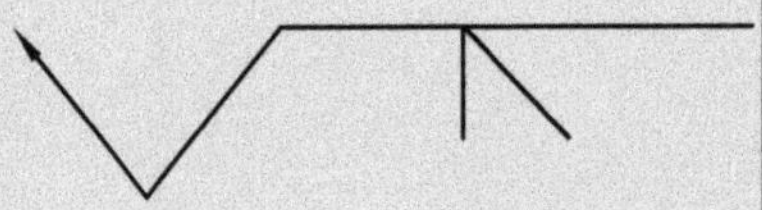	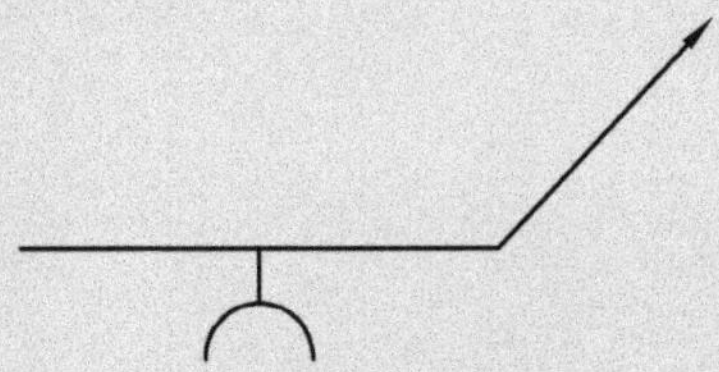
OTHER SIDE	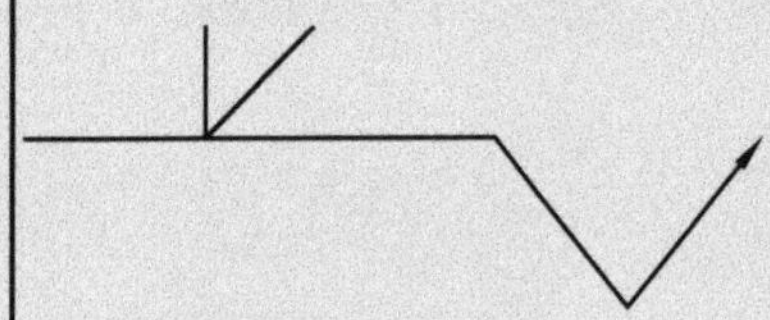	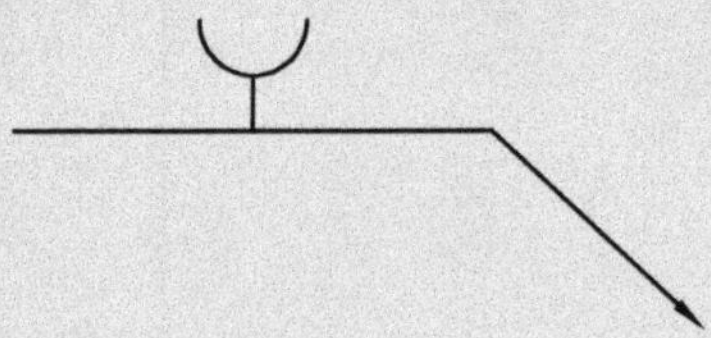
BOTH SIDES		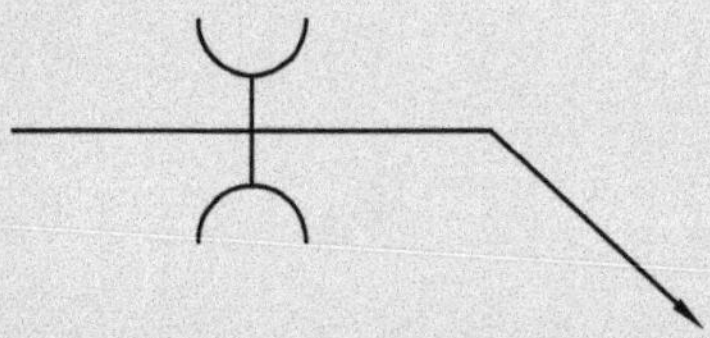
NO ARROW SIDE OR OTHER SIDE SIGNIFICANCE	NOT USED	NOT USED

REFERENCE 8-7

GROVE WELDING SYMBOLS AND THEIR LOCATION SIGNIFICANCE

LOCATION SIGNIFICANCE	J	FLARE-V
ARROW SIDE		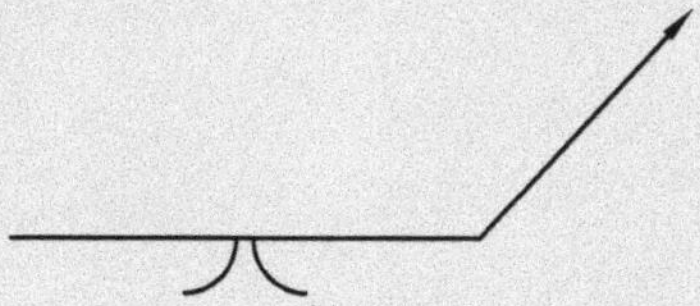
OTHER SIDE		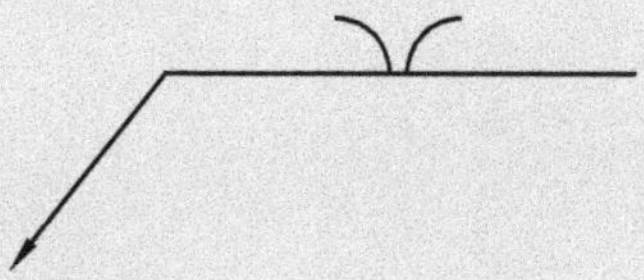
BOTH SIDES	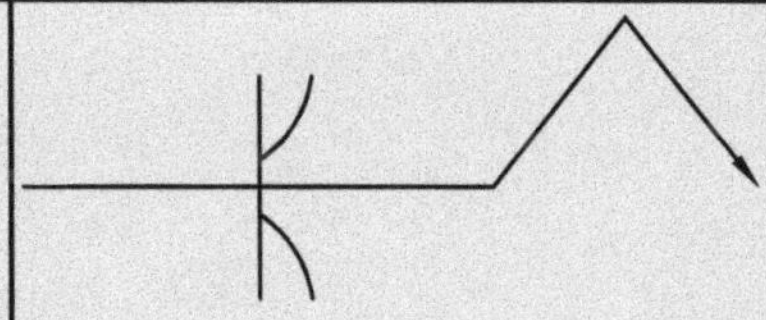	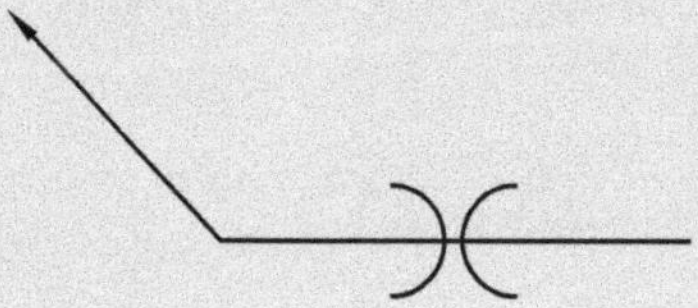
NO ARROW SIDE OR OTHER SIDE SIGNIFICANCE	NOT USED	NOT USED

REFERENCE 8-8

GROVE WELDING SYMBOLS AND THEIR LOCATION SIGNIFICANCE

LOCATION SIGNIFICANCE	FLARE-BEVEL	SCART FOR BRAZED JOINT
ARROW SIDE	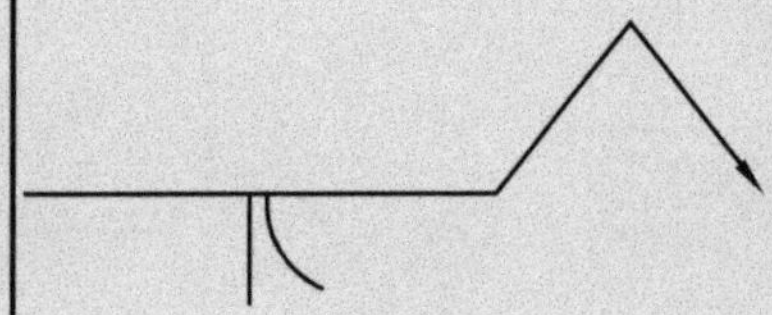	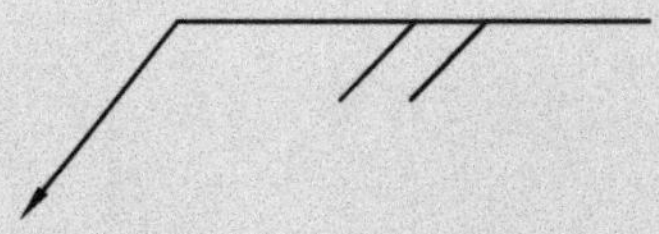
OTHER SIDE		
BOTH SIDES	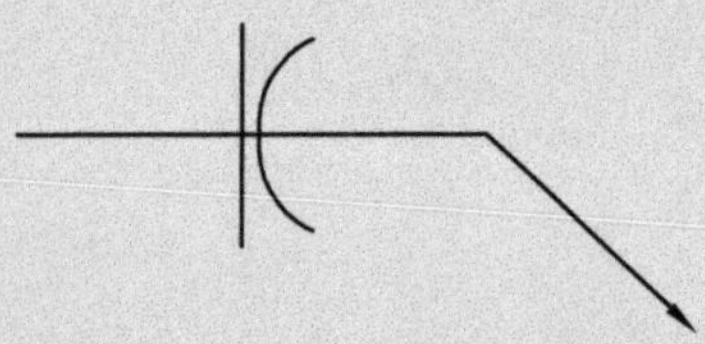	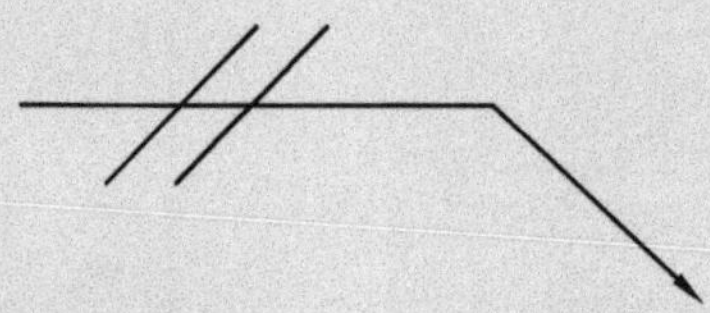
NO ARROW SIDE OR OTHER SIDE SIGNIFICANCE	NOT USED	NOT USED

REFERENCE 8-9

SUPPLEMENTARY WELDING SYMBOLS

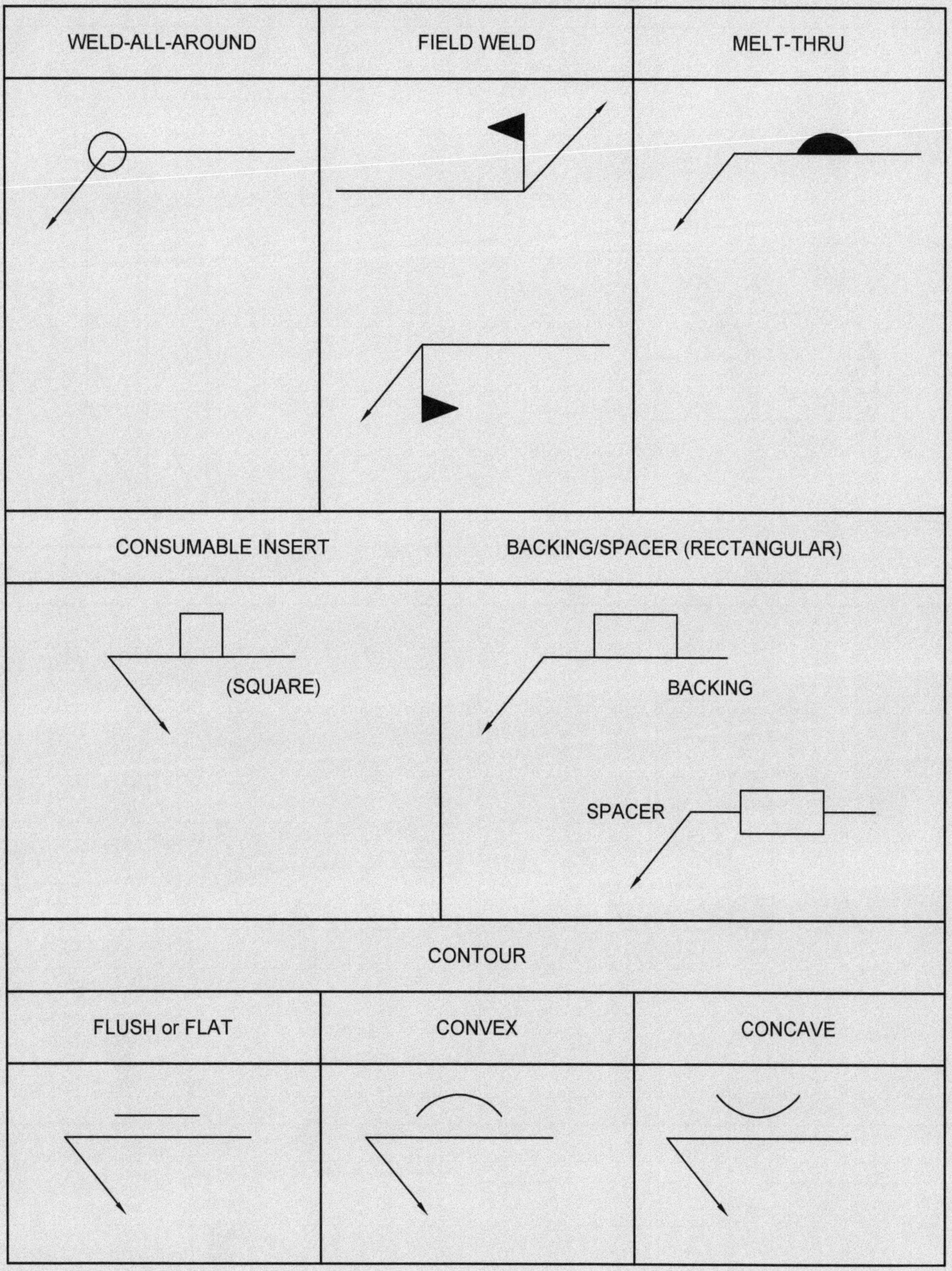

REFERENCE 8-10

BASIC JOINTS IDENTIFICATION OF ARROW SIDE AND OTHER SIDE JOINTS

BUTT JOINT

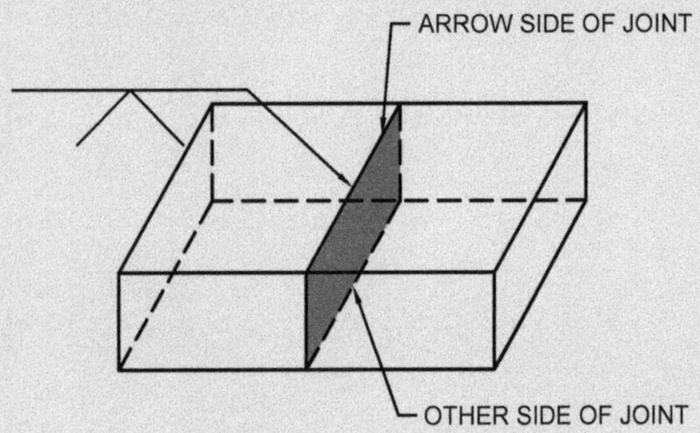

CORNER JOINT

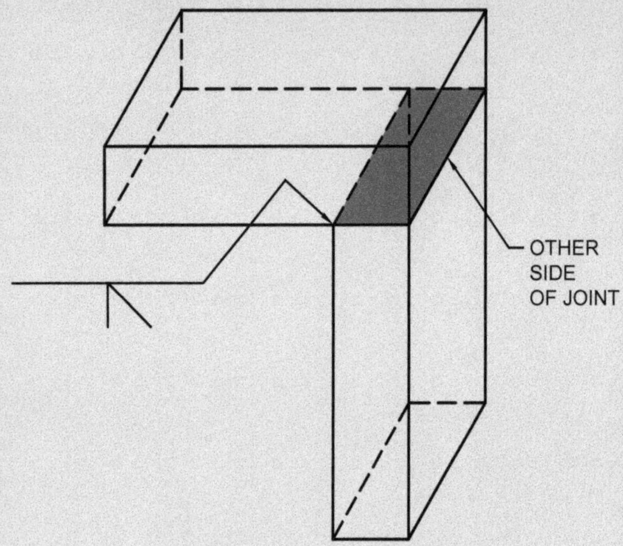

T-JOINT

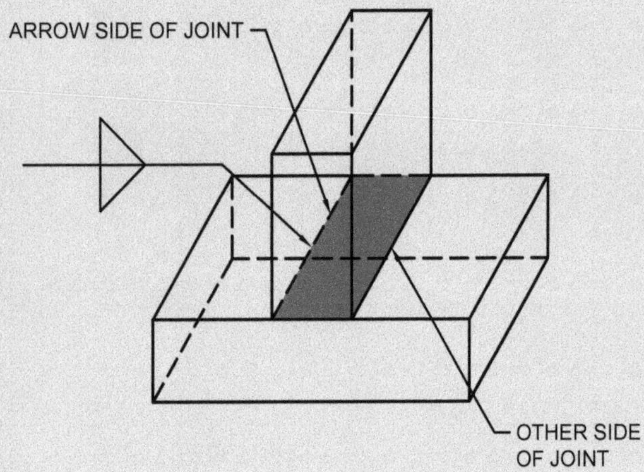

LAP JOINT

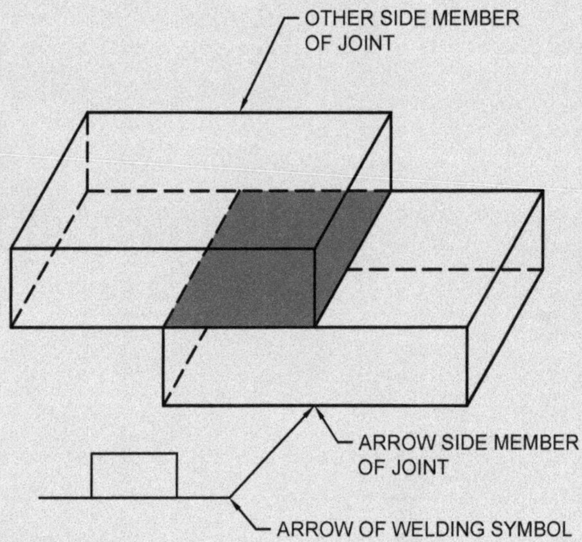

EDGE JOINT

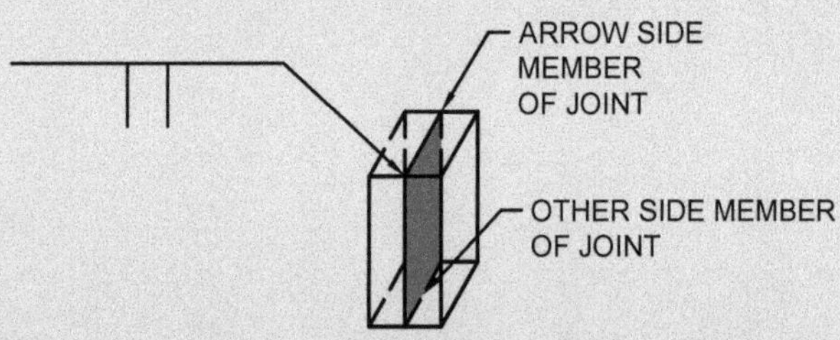

LOCATION OF ELEMENTS OF A WELDING SYMBOL

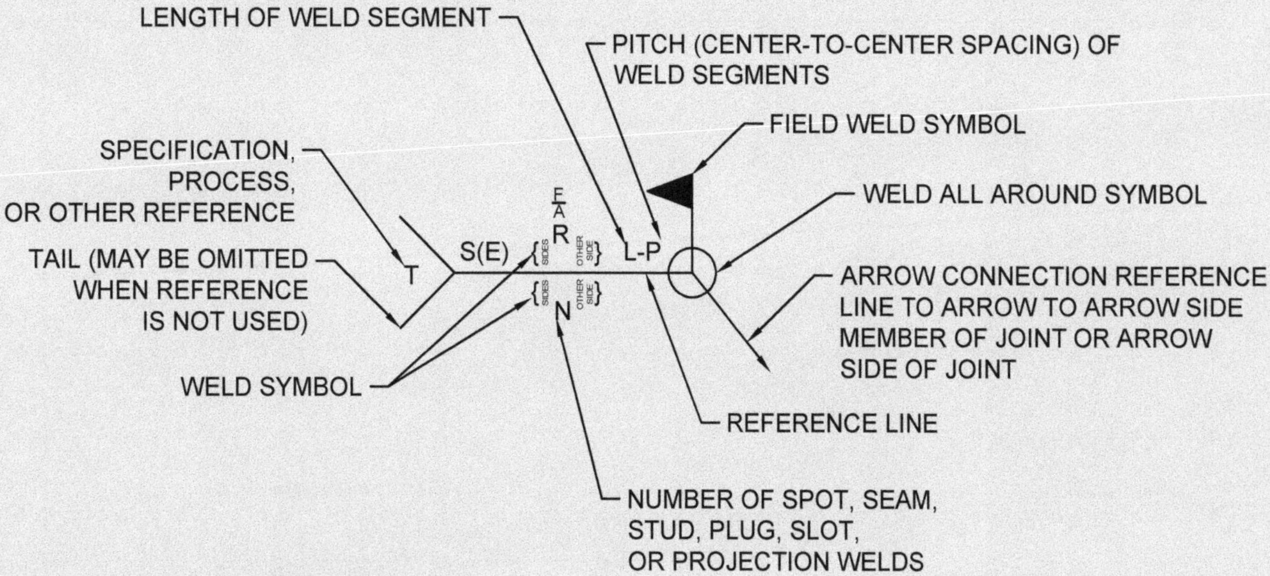

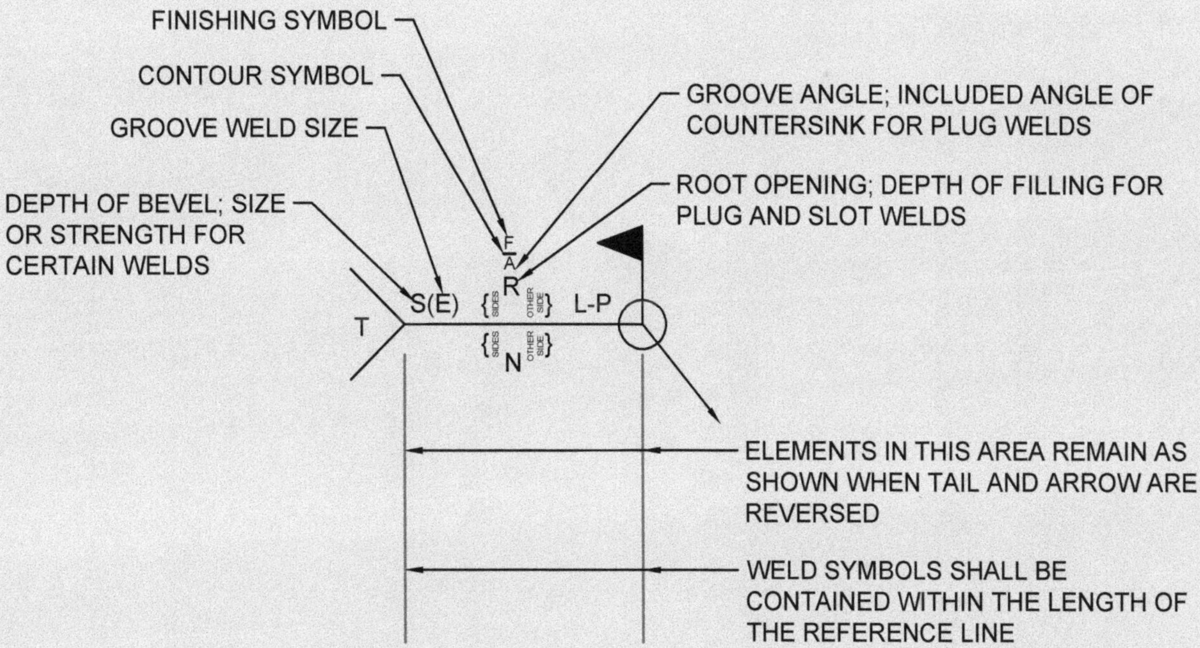

REFERENCE 8-12

TYPICAL WELDING SYMBOLS

DOUBLE-FILLET WELDING SYMBOL

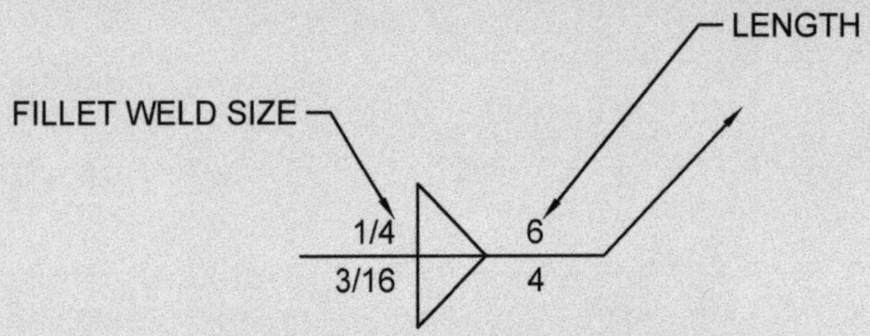

OMISSION OF LENGTH INDICATES THAT WELD EXTENDS BETWEEN ABRUPT CHANGES IN DIRECTION OR AS DIMENSIONED

PLUG WELDING SYMBOL

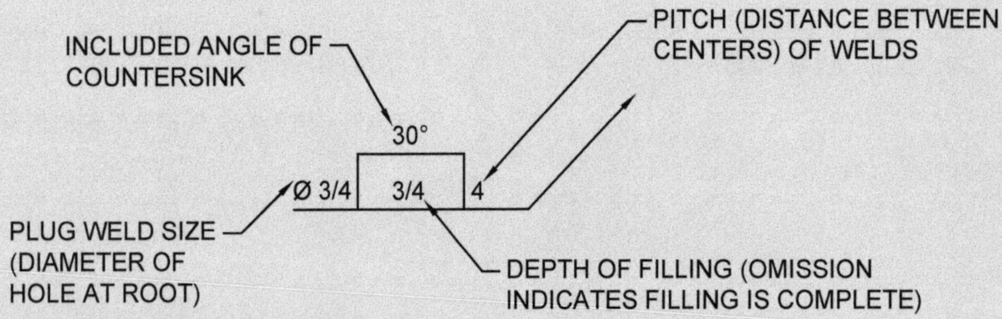

SPOT WELDING SYMBOL

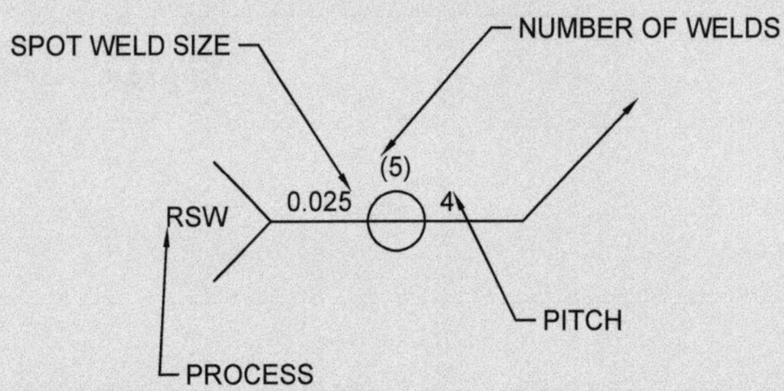

REFERENCE 8-13

TYPICAL WELDING SYMBOLS

CHAIN INTERMITTENT FILLET WELDING SYMBOL

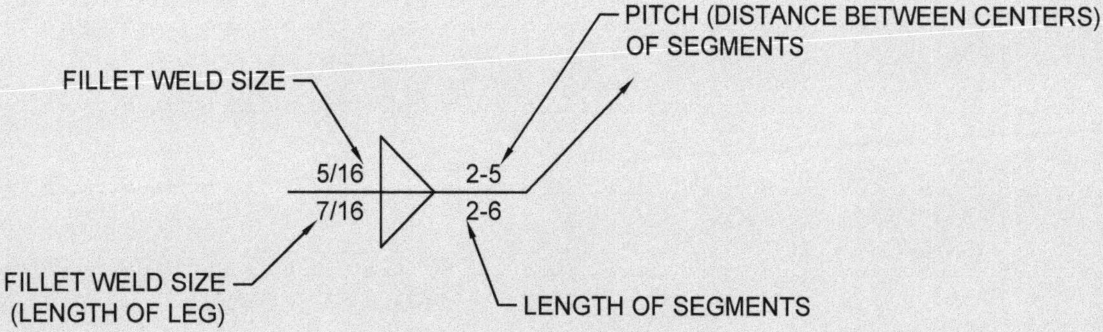

BACK WELDING SYMBOL

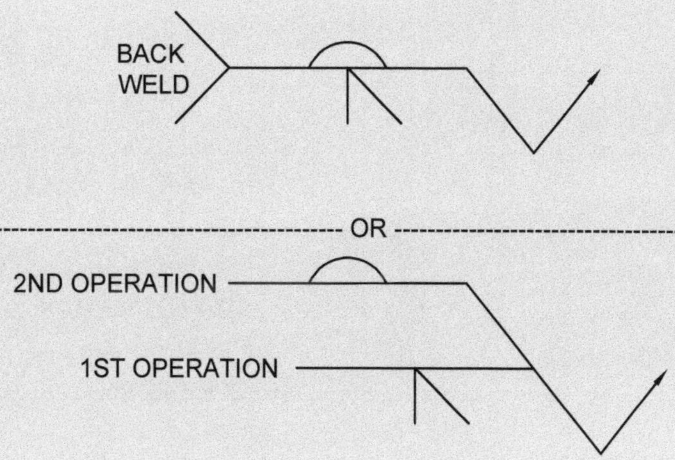

STUD WELDING SYMBOL

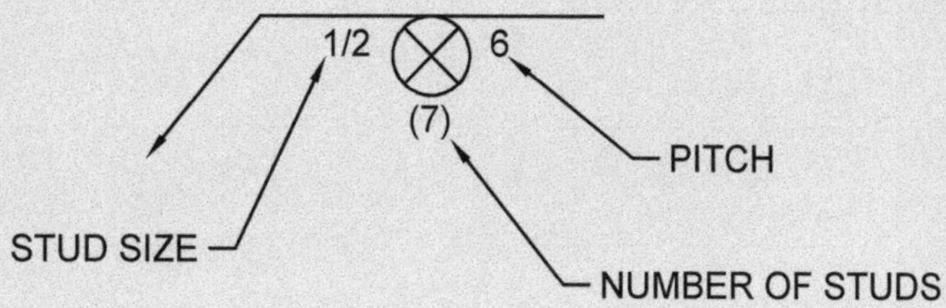

TYPICAL WELDING SYMBOLS

STAGGERED INTERMITTENT FILLET WELDING SYMBOL

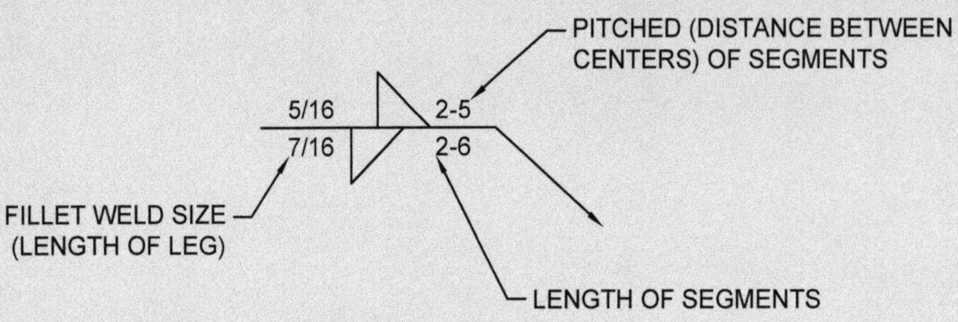

BACKING WELDING SYMBOL

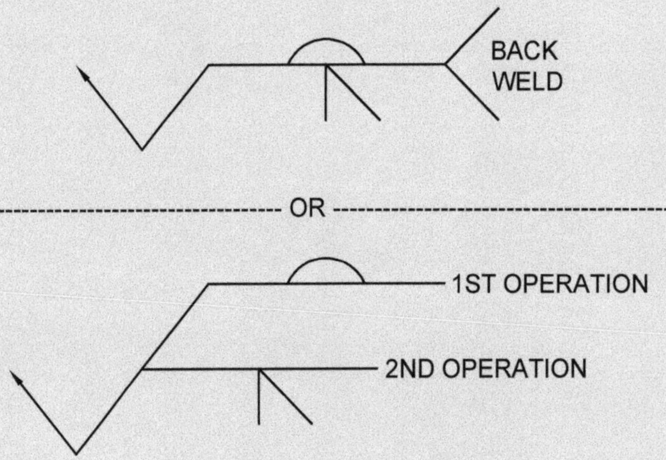

SEAM WELDING SYMBOL

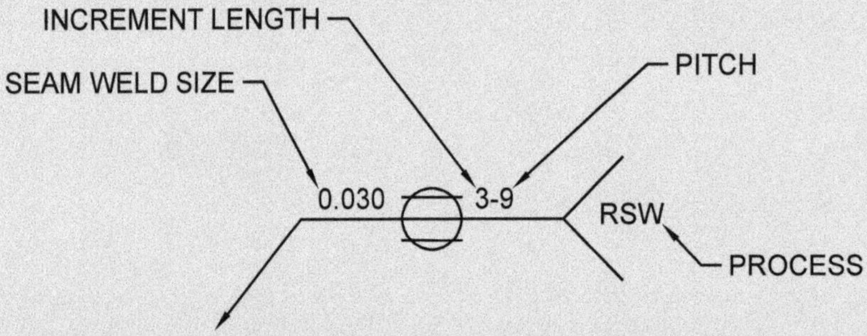

REFERENCE 8-15

TYPICAL WELDING SYMBOLS

SQUARE-GROOVE WELDING SYMBOL

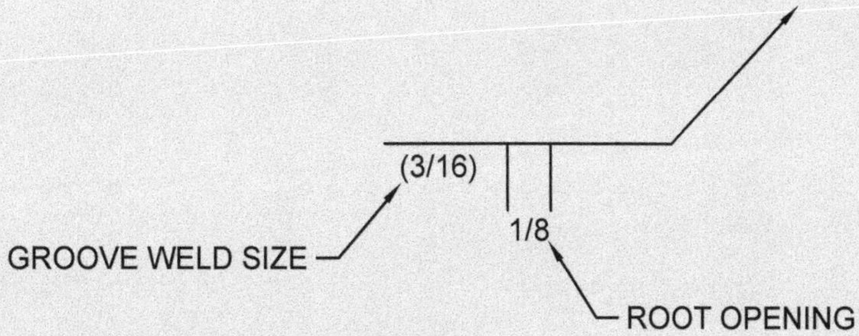

SYMBOL WITH BACKGOUGING

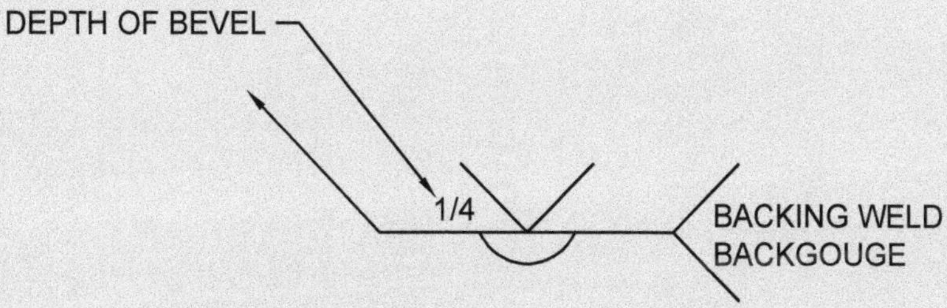

MULTIPLE REFERENCE LINES

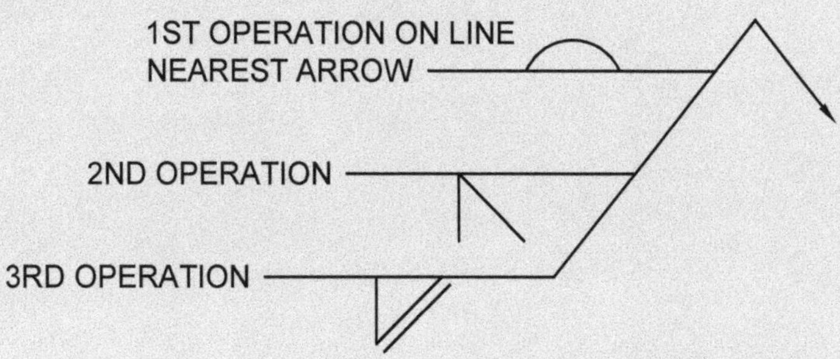

TYPICAL WELDING SYMBOLS

V-GROOVE WELDING SYMBOL

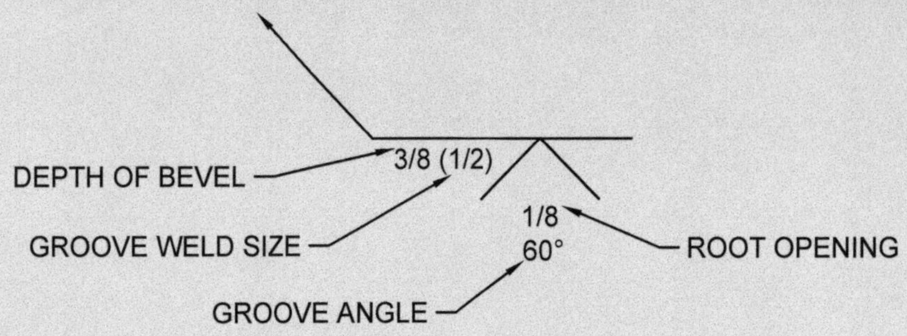

FLARE-V-GROOVE WELDING SYMBOL

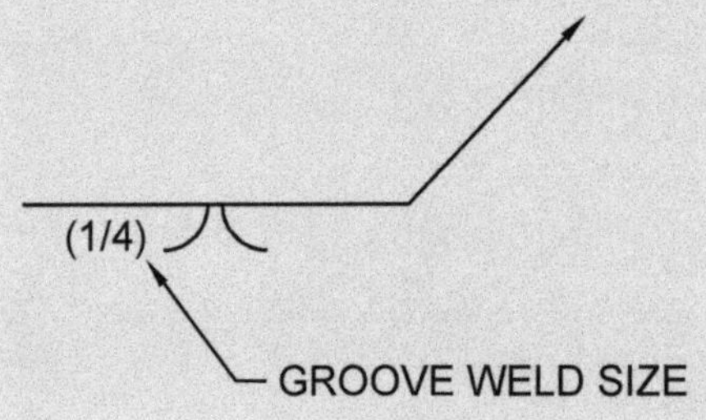

COMPLETE JOINT PENETRATION

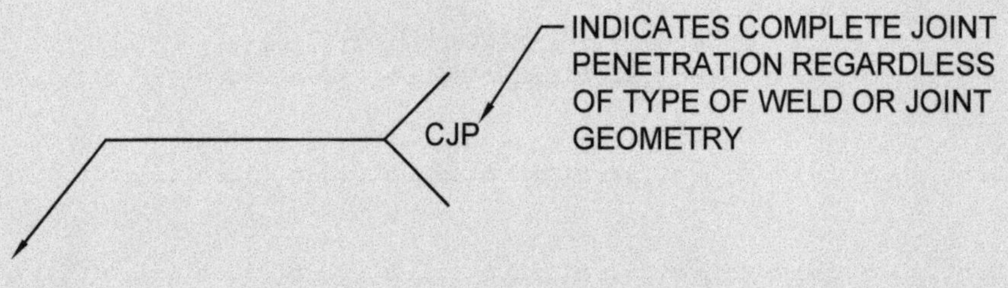

REFERENCE 8-17

TYPICAL WELDING SYMBOLS

DOUBLE-BEVEL-GROOVE WELDING SYMBOL

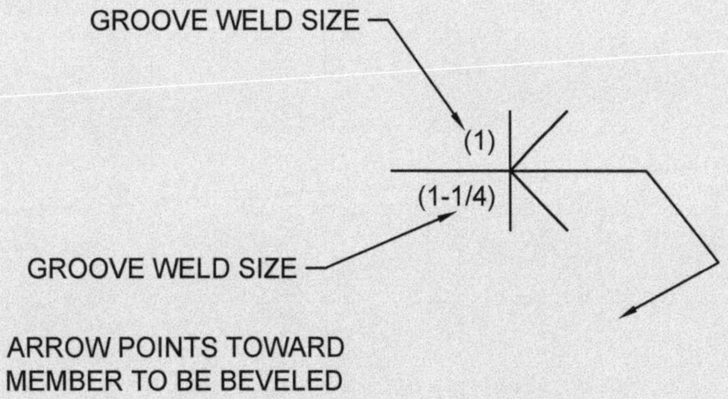

FLARE-BEVEL-GROOVE WELDING SYMBOL

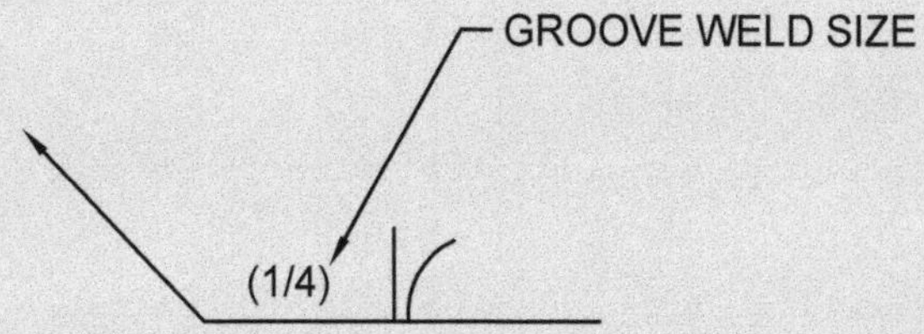

EDGE WELDING SYMBOL

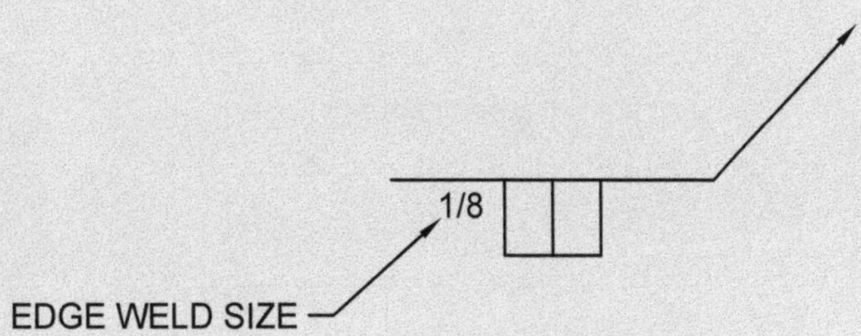

REFERENCE 8-18

TYPICAL WELDING SYMBOLS

FLASH or UPSET WELDING SYMBOL

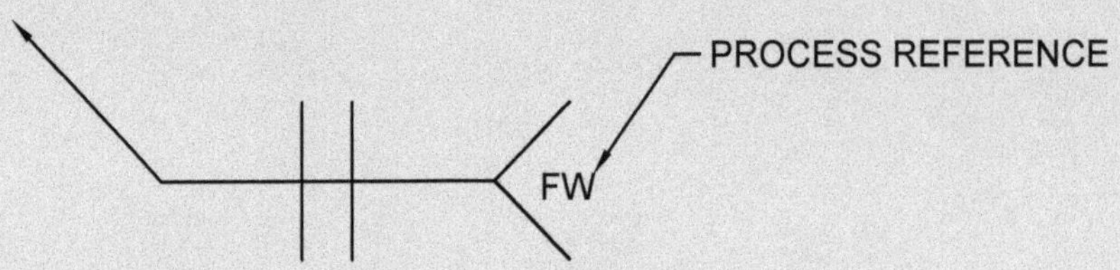

JOINT WITH SPACER

WITH MODIFIED GROOVE WELD SYMBOL

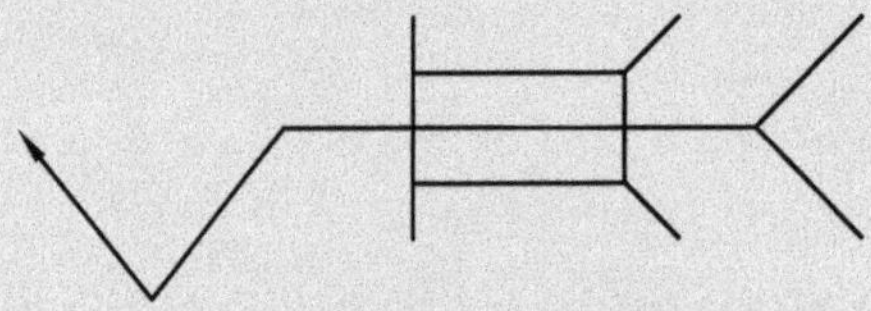

DOUBLE-BEVEL GROOVE

MELT-THRU SYMBOL

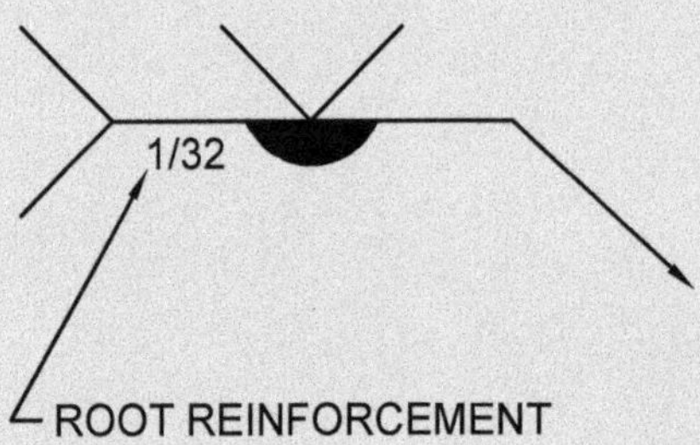

TYPICAL WELDING SYMBOLS

FLUSH CONTOUR SYMBOL

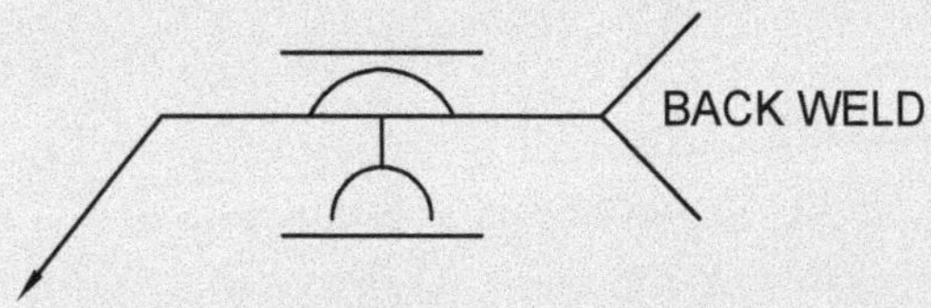

JOINT WITH BACKING

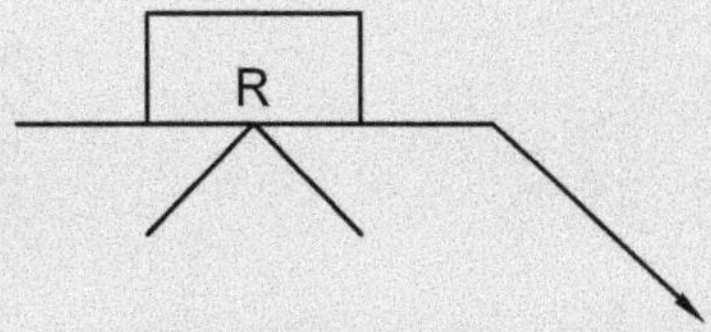

'R' INDICATES BACKING
REMOVED AFTER WELDING

CONVEX CONTOUR SYMBOL

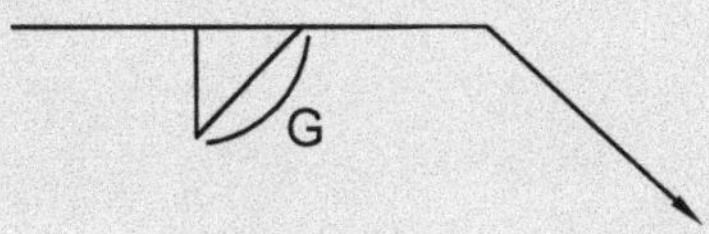

STANDARD WELDING SYMBOLS (BASED ON AWS)

	FILLET	PLUG/SLOT	SPOT
ARROW SIDE	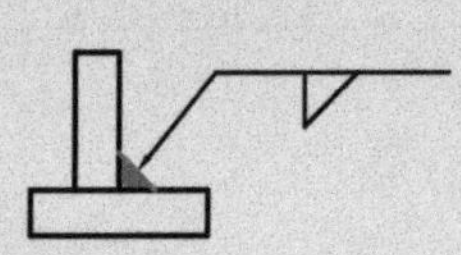	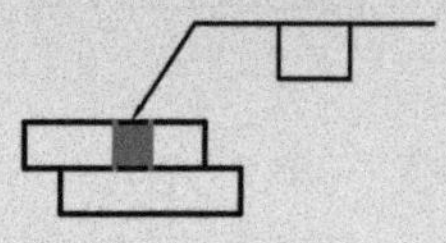	LAP-JOINT 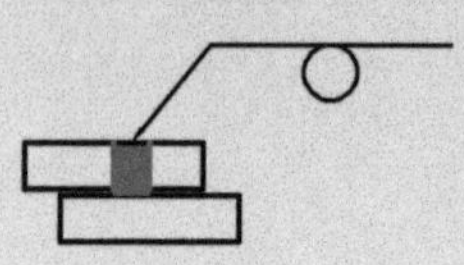
OTHER SIDE	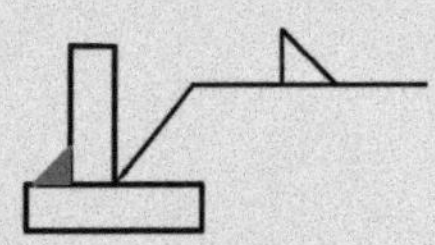	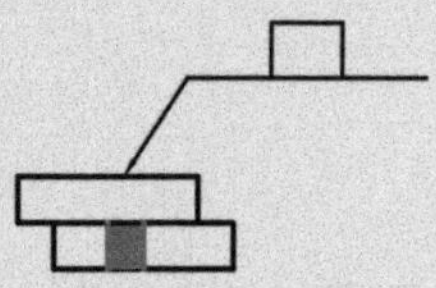	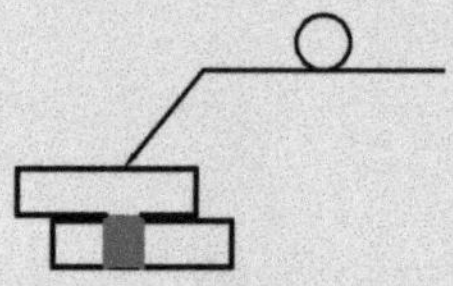
BOTH SIDES	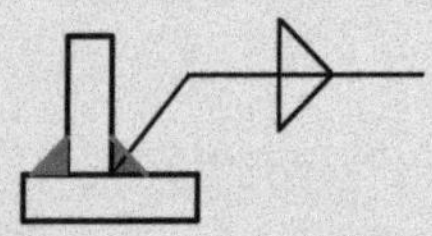	NOT USED	NOT USED
NO SIDE USED	NOT USED	NOT USED	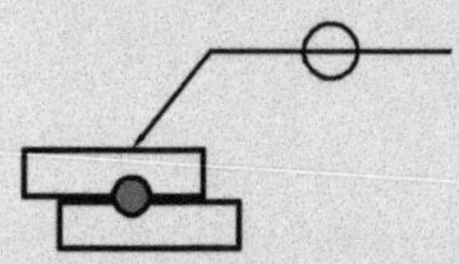

REFERENCE 8-21

STANDARD WELDING SYMBOLS (BASED ON AWS)

	STUD	SEAM	BACKING RUN
ARROW SIDE	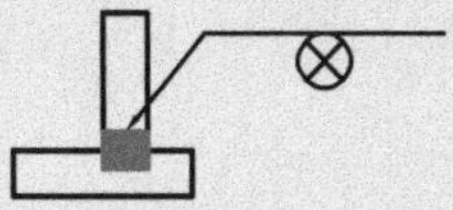	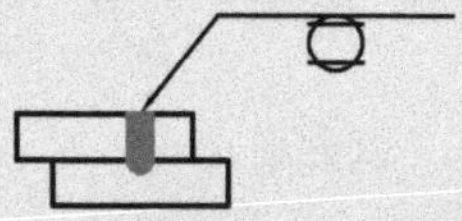	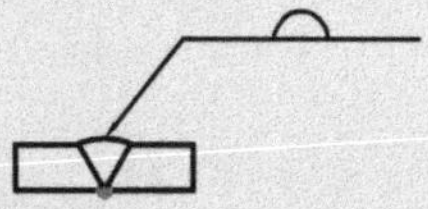
OTHER SIDE	NOT USED	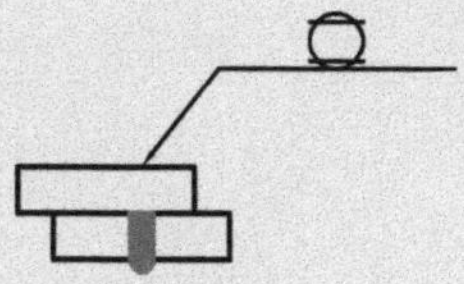	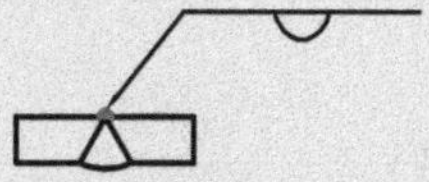
BOTH SIDES	NOT USED	NOT USED	NOT USED
NO SIDE USED	NOT USED	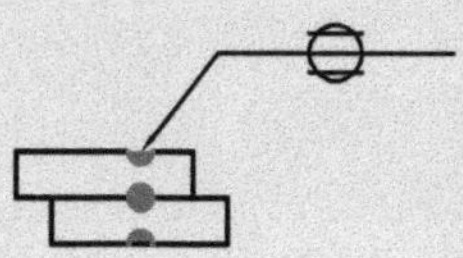	NOT USED

REFERENCE 8-22

STANDARD WELDING SYMBOLS (BASED ON AWS)

	SURFACING	EDGE	FLANGE EDGE
ARROW SIDE	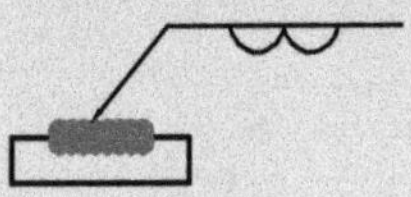	EDGE-JOINT 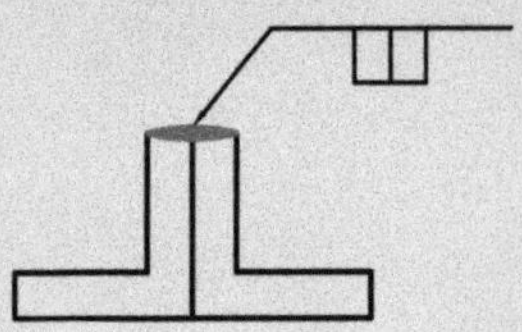	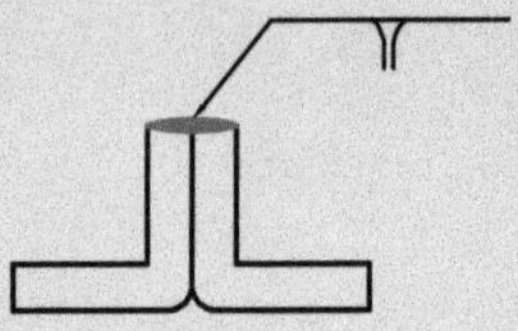
OTHER SIDE	NOT USED		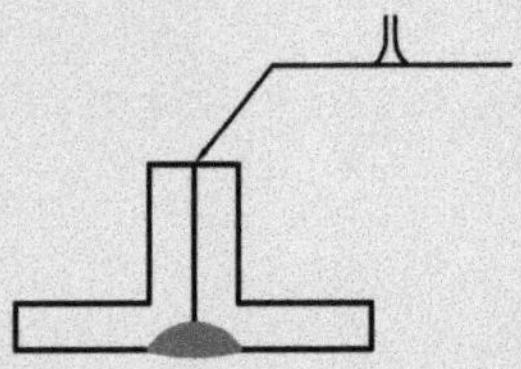
BOTH SIDES	NOT USED		NOT USED
NO SIDE USED	NOT USED	NOT USED	NOT USED

REFERENCE 8-23

STANDARD WELDING SYMBOLS (BASED ON AWS)

	SQUARE-GROOVE	V BUTT-JOINT	BEVEL CORNER-JOINT
ARROW SIDE	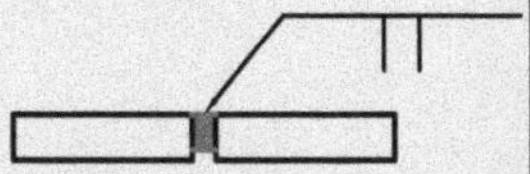	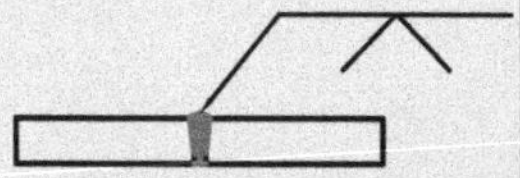	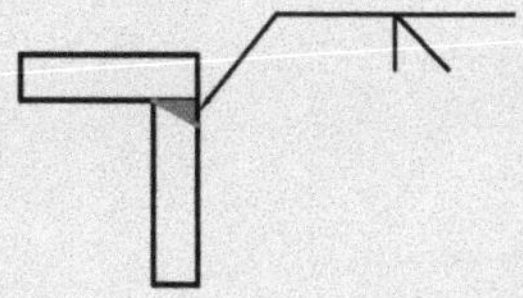
OTHER SIDE	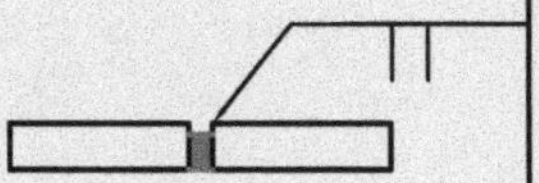	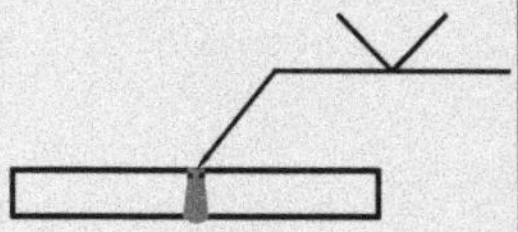	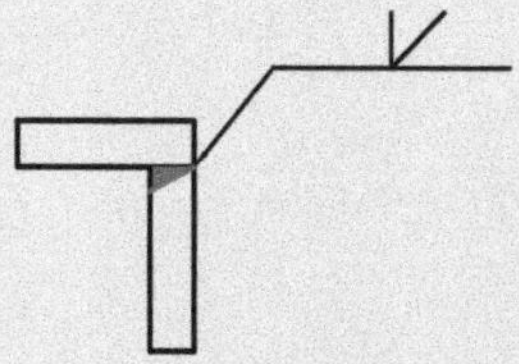
BOTH SIDES	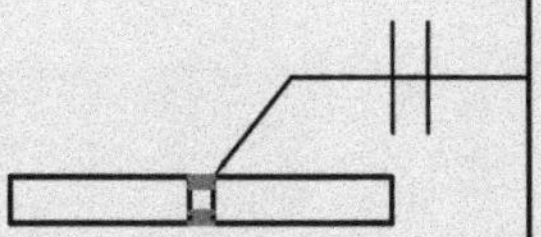	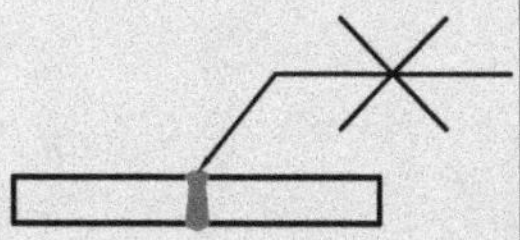	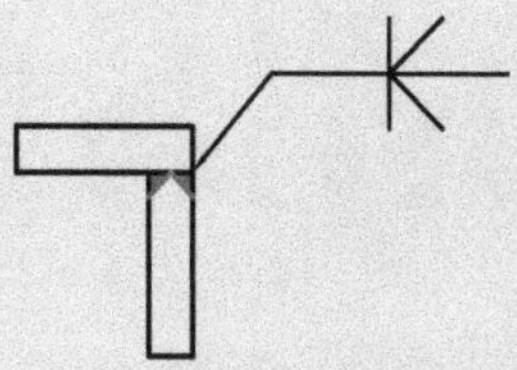
NO SIDE USED	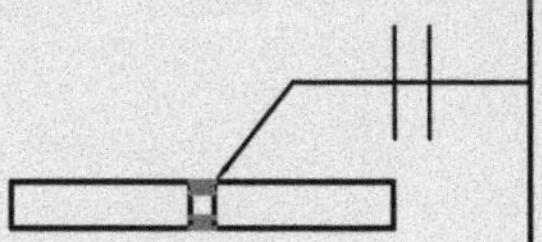	NOT USED	NOT USED

	U
ARROW SIDE	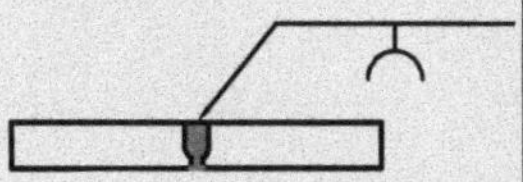
OTHER SIDE	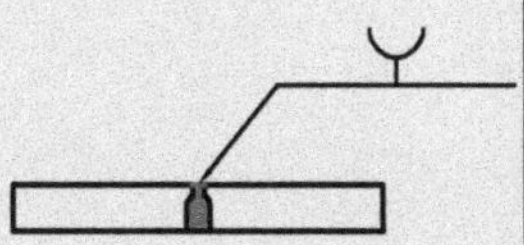

MISCELLANEOUS CAD DETAILS

REFERENCE 9-1

4" AND 6" CMU BLOCK DIMENSIONS

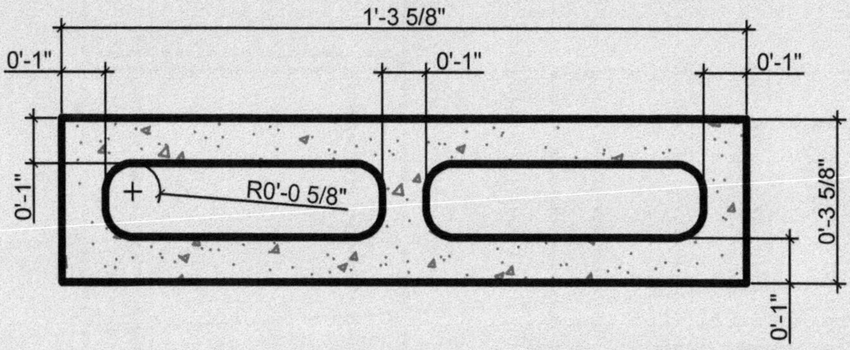

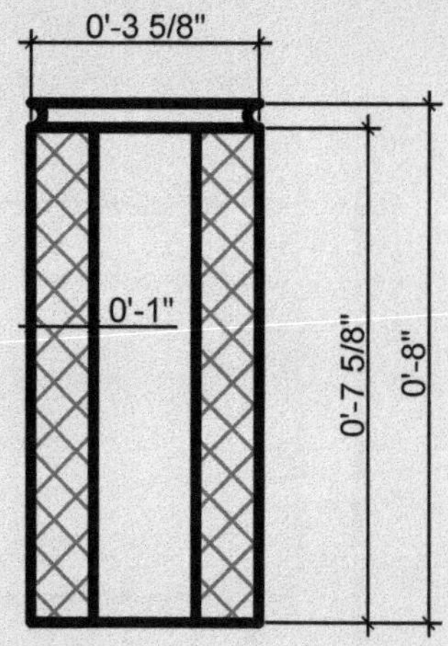

1 4-INCH CMU

3" = 1'-0"

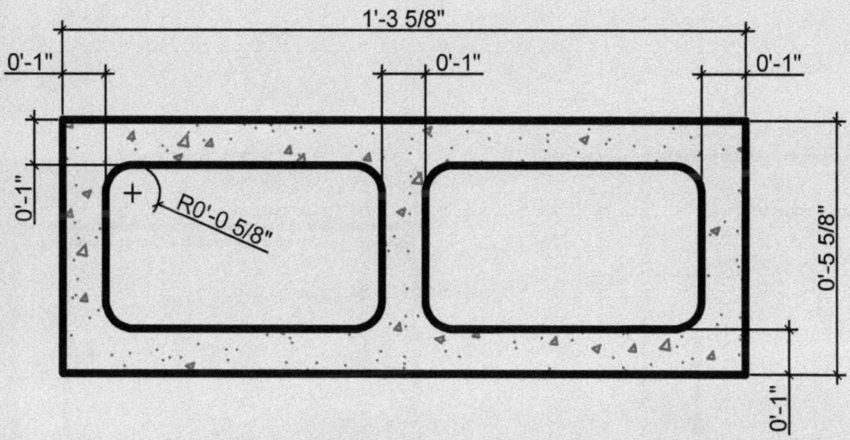

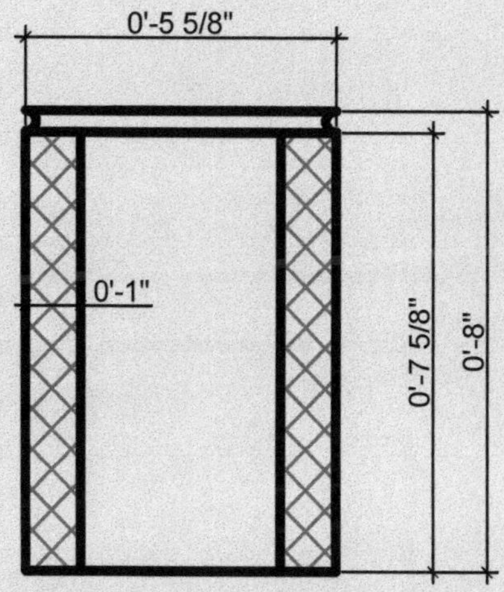

1 6-INCH CMU

3" = 1'-0"

REFERENCE 9-2

8" AND 10" CMU BLOCK DIMENSIONS

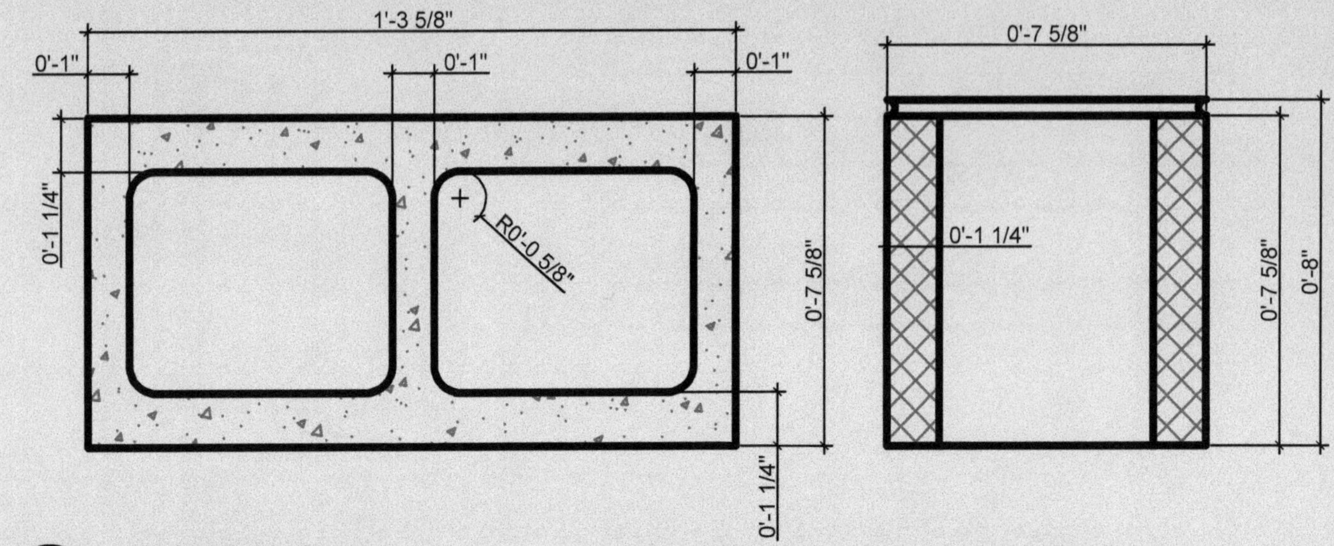

1 8-INCH CMU

3" = 1'-0"

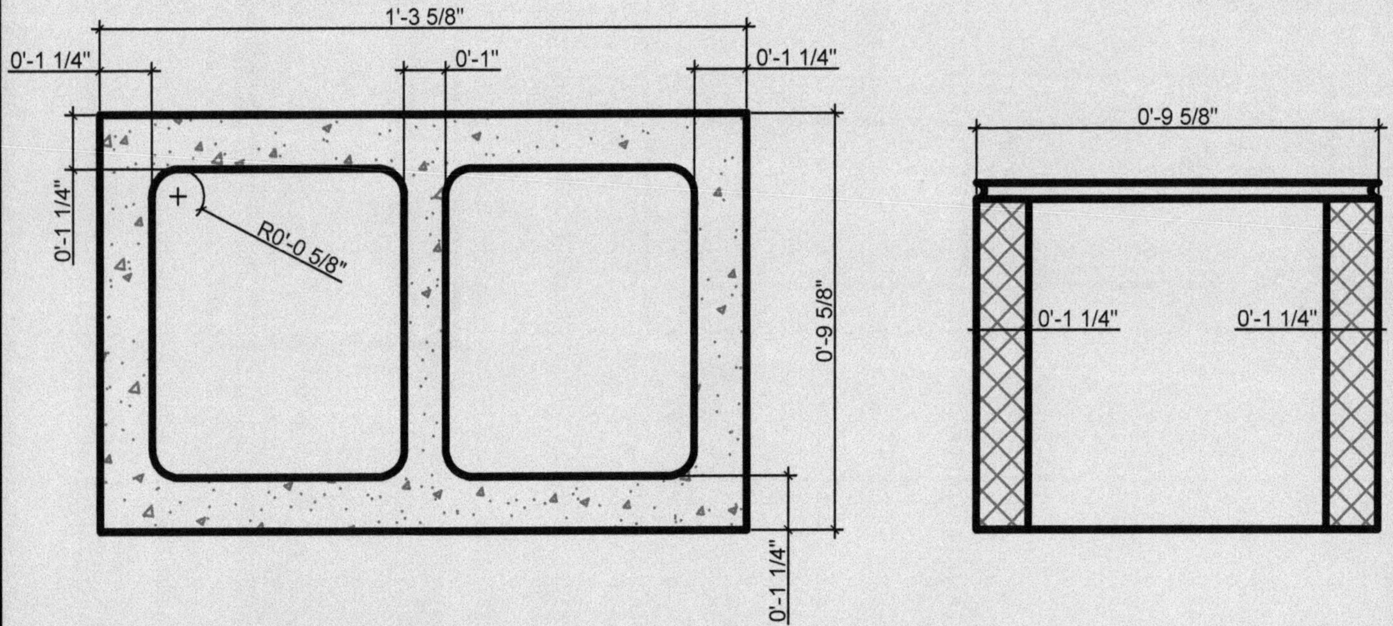

1 10-INCH CMU

3" = 1'-0"

REFERENCE 9-3

12" CMU BLOCK DIMENSIONS

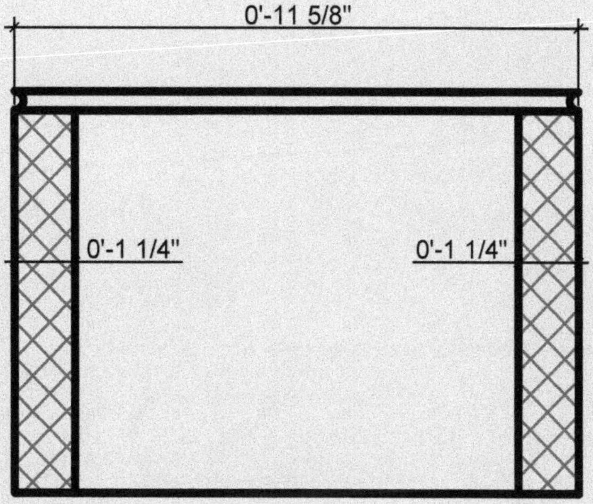

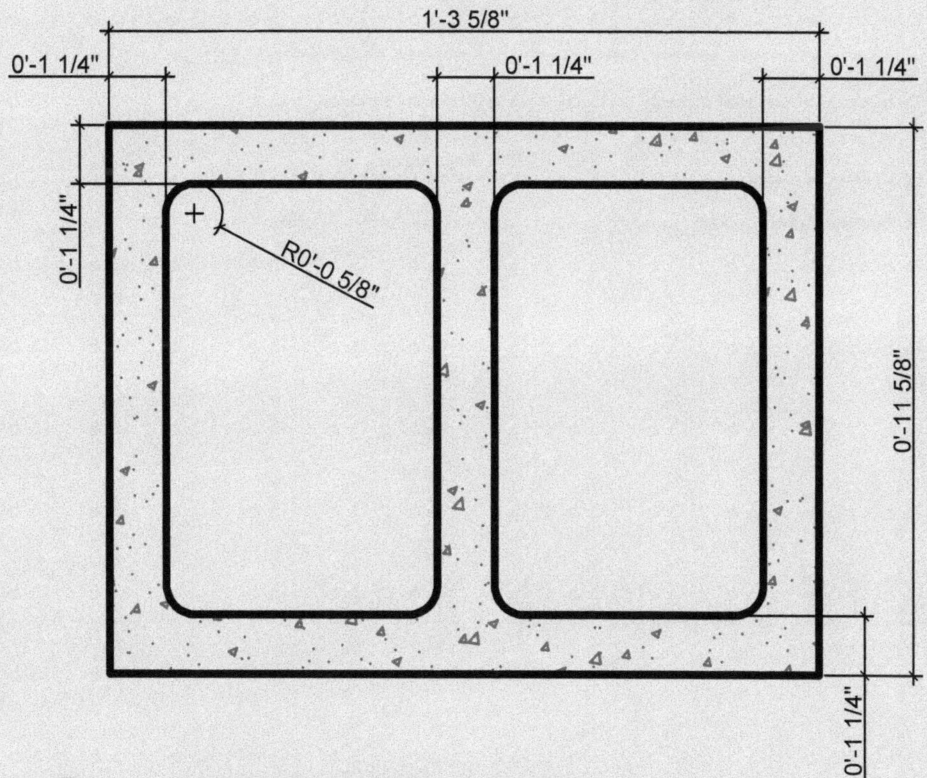

1 12-INCH CMU

3" = 1'-0"

TYPES OF MASONRY UNITS

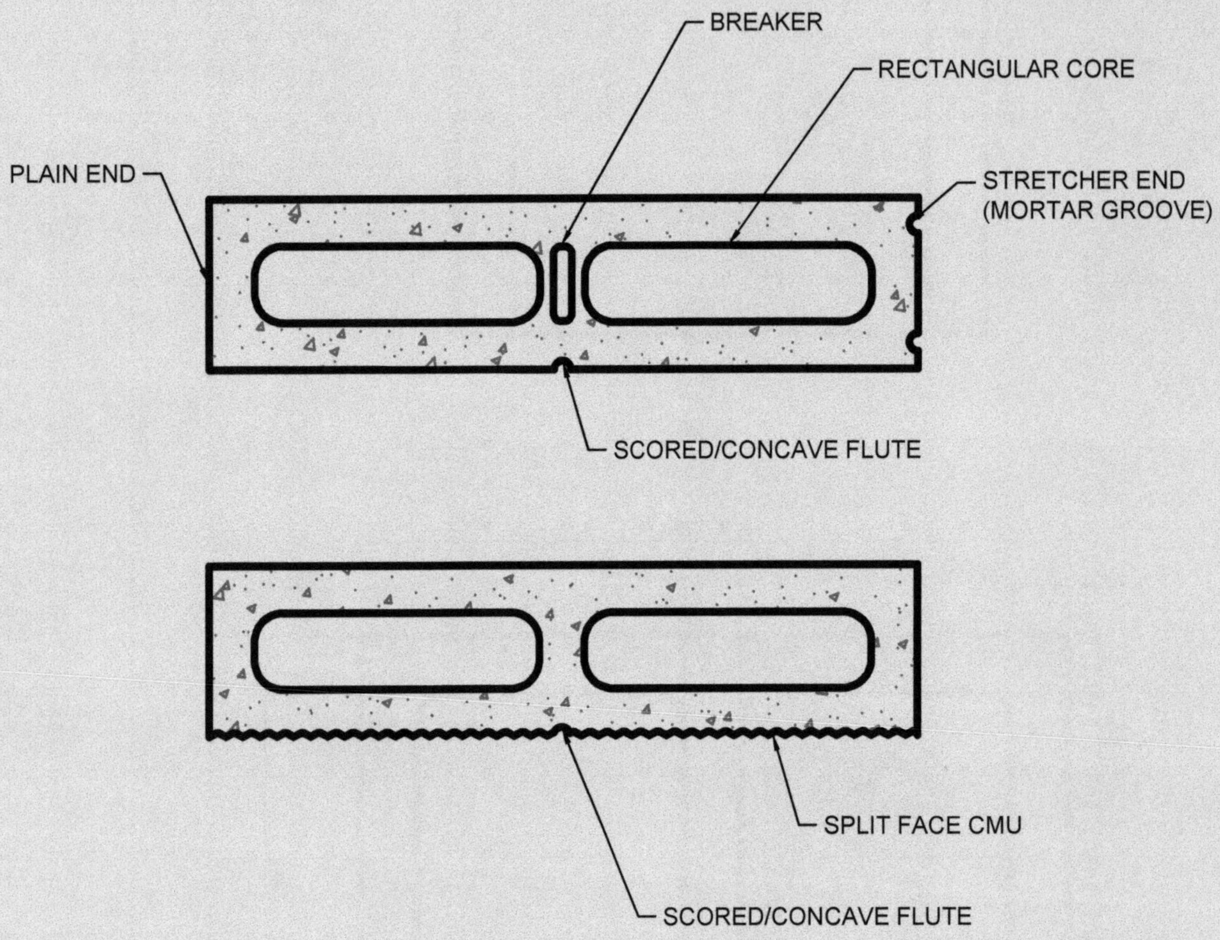

1 TYPES OF MASONRY UNITS

3" = 1'-0"

REFERENCE 9-5

6" AND 8" CMU LINTELS AND KNOCK OUT BOND BEAMS

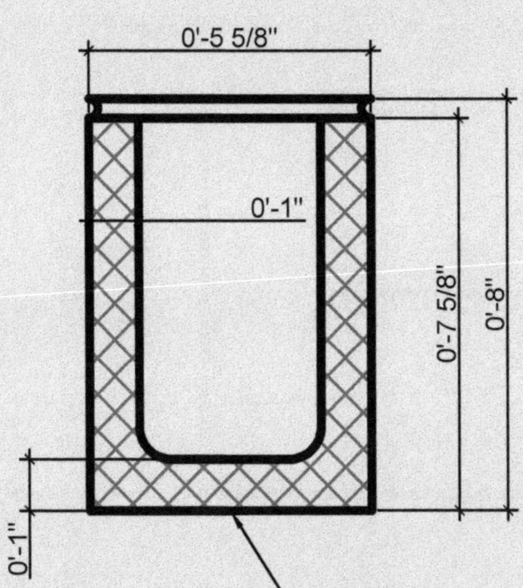

6" SOLID BOTTOM LINTEL

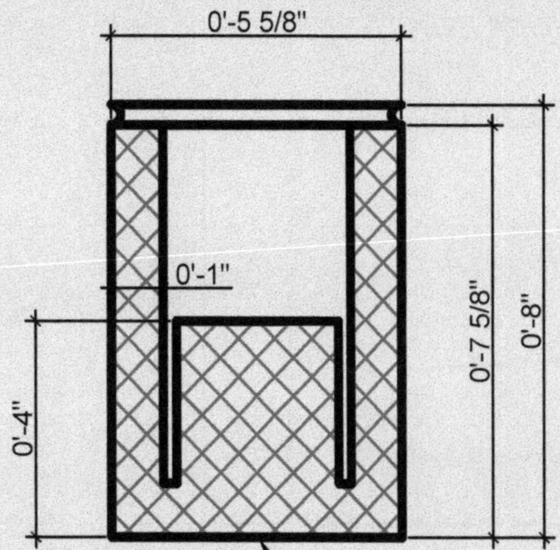

6" KNOCK OUT BOND BEAM RESTRICTED OPEN BOTTOM LINTEL

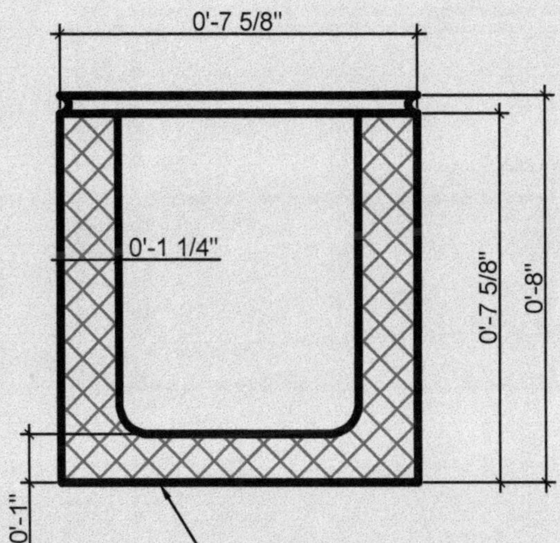

8" SOLID BOTTOM LINTEL

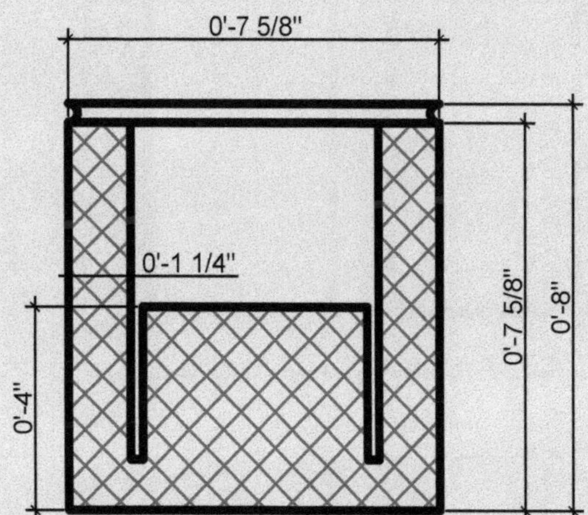

8" KNOCK OUT BOND BEAM RESTRICTED OPEN BOTTOM LINTEL

1 6-INCH & 8-INCH LINTELS

3" = 1'-0"

REFERENCE 9-6

10" AND 12" CMU LINTELS AND KNOCK OUT BOND BEAMS

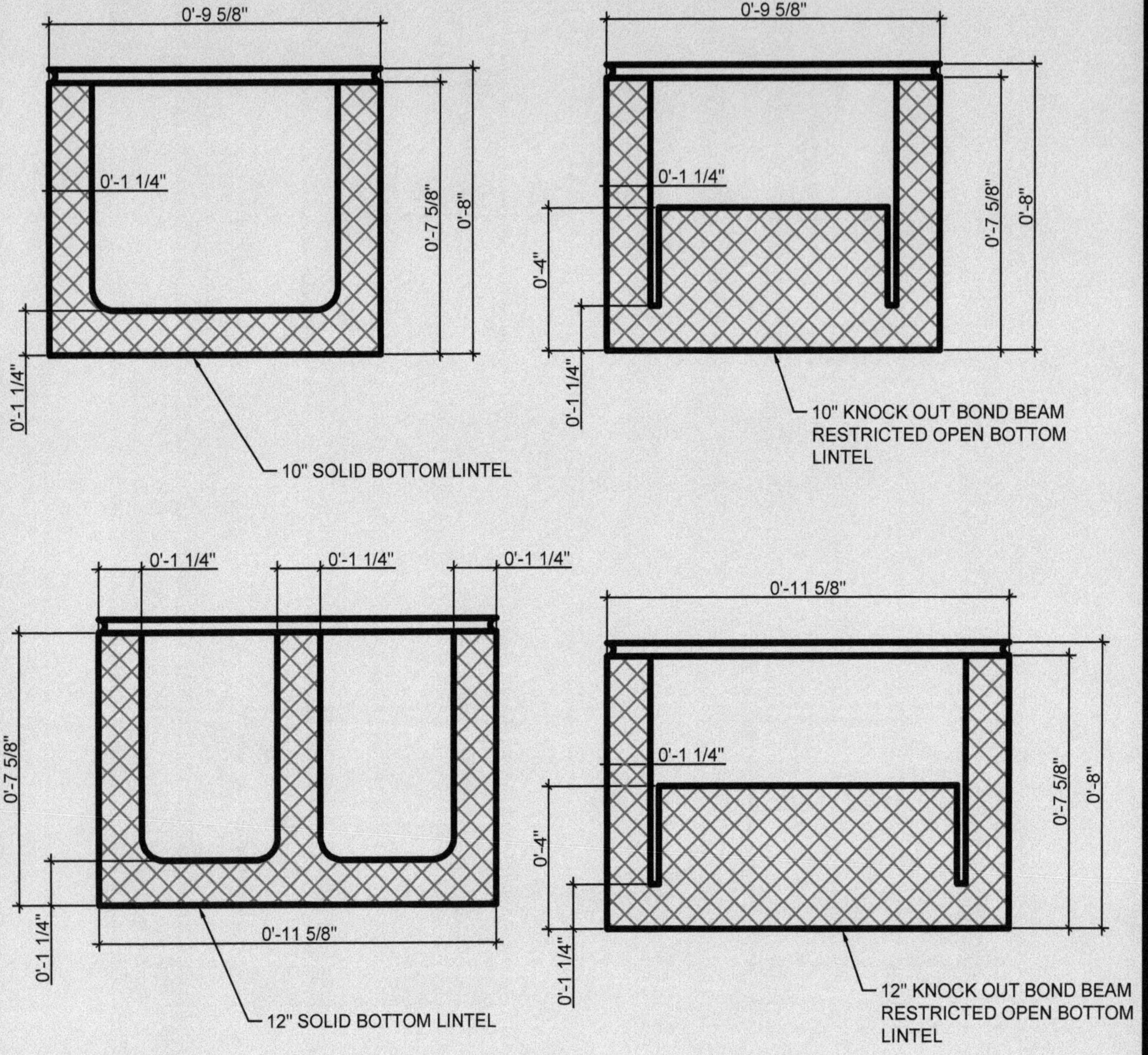

1 10-INCH & 12-INCH LINTELS

3" = 1'-0"

COMMON BRICK DIMENSIONS

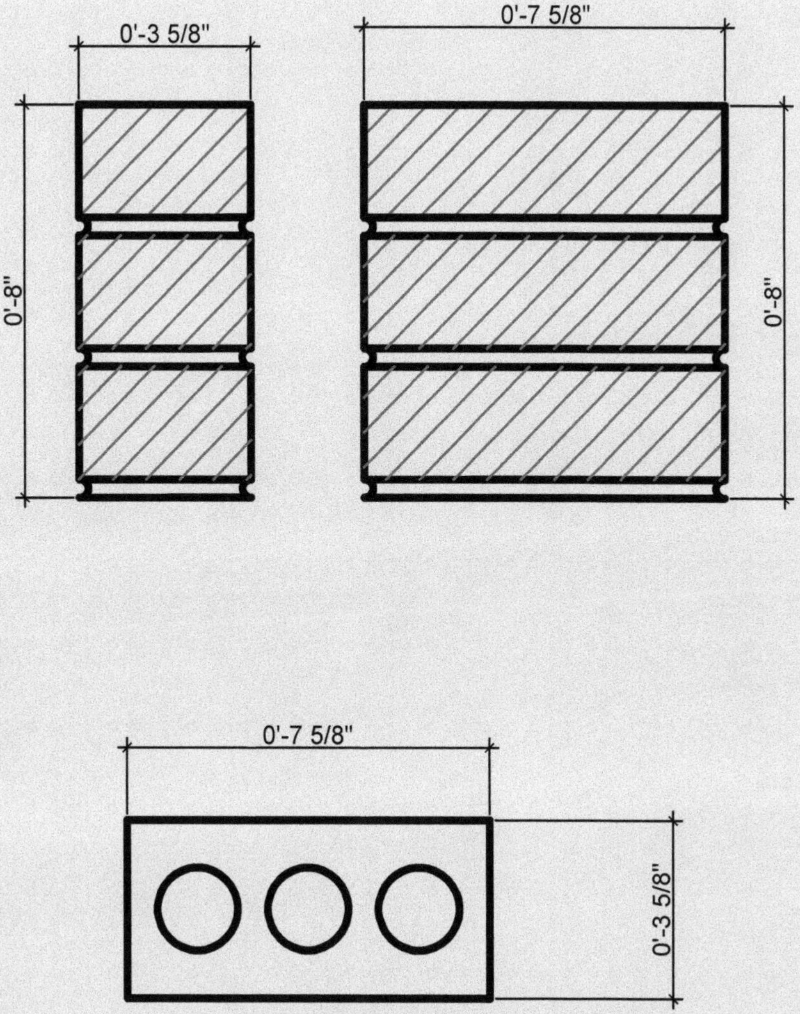

1 COMMON BRICK DIMENSIONS

3" = 1'-0"

BRICK BACK FILLED WITH METAL STUDS WALL FOUNDATION DETAIL

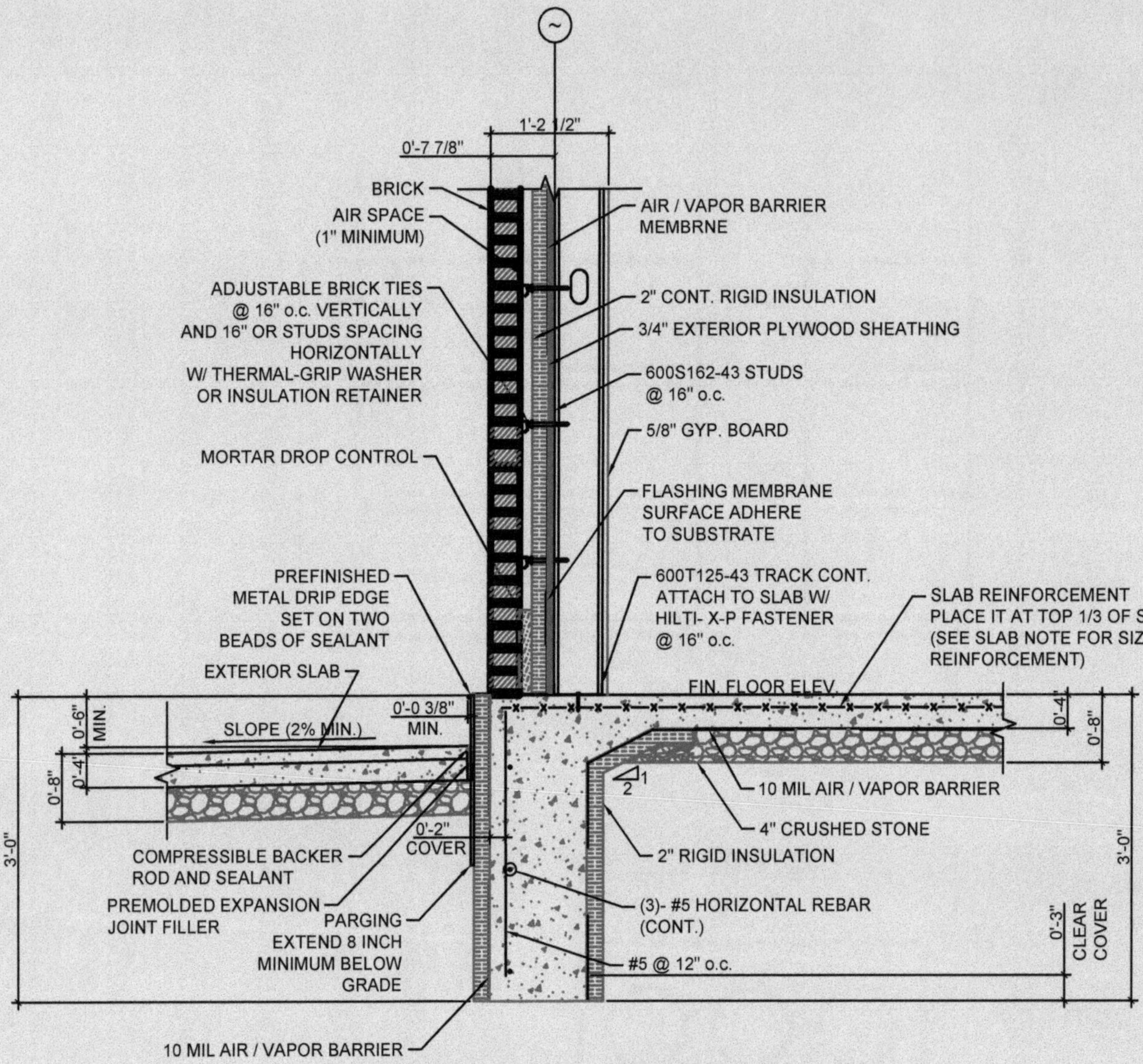

1 FOUNDATION DETAIL

3/4" = 1'-0"

BRICK BACK FILLED WITH METAL STUDS WALL FOUNDATION DETAIL

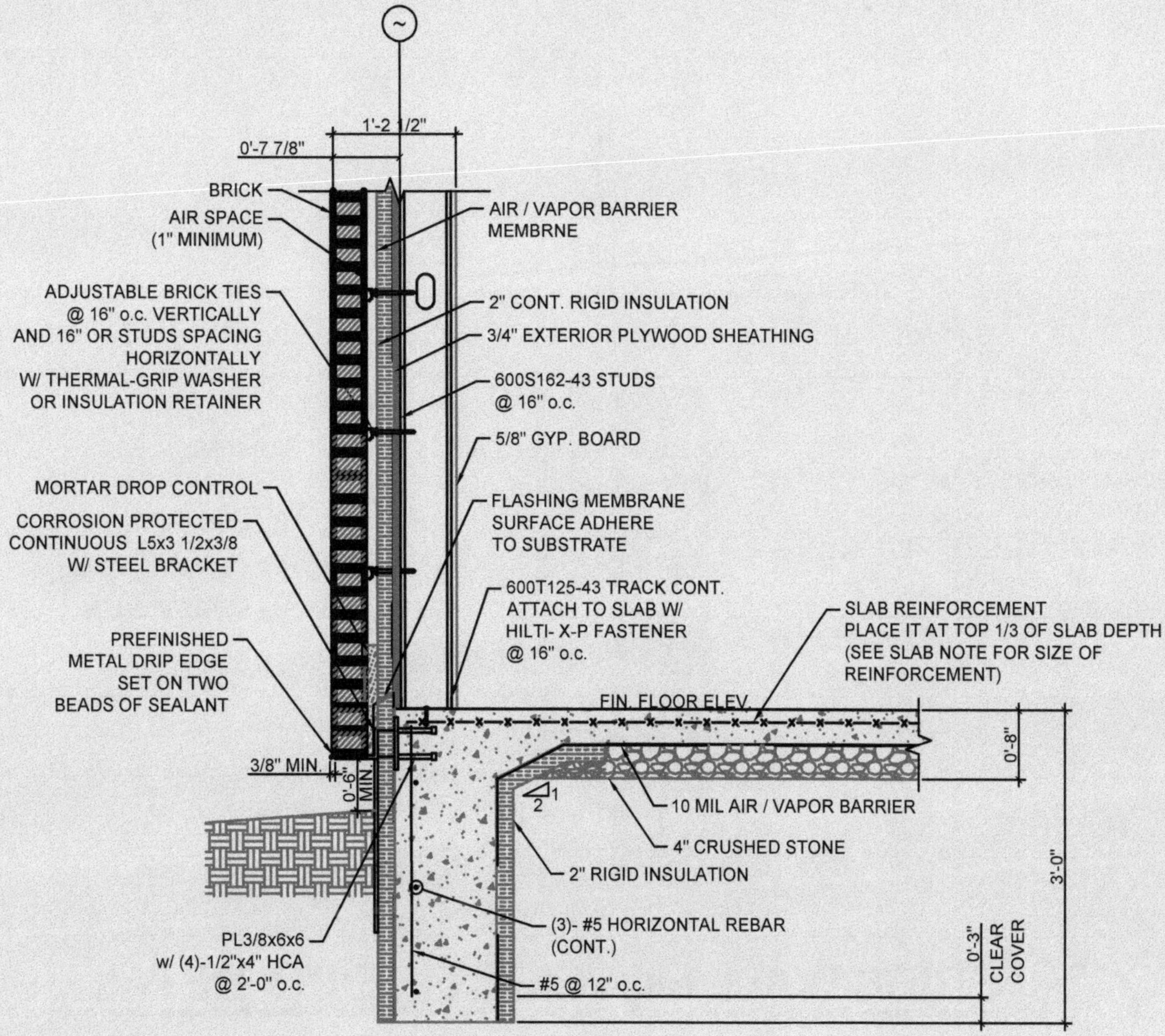

1 ALTERNATIVE FOUNDATION DETAIL
(SHELF ANGLE WITH SOIL) 3/4" = 1'-0"

REFERENCE 9-10

BRICK WALL EXPANSION JOINT

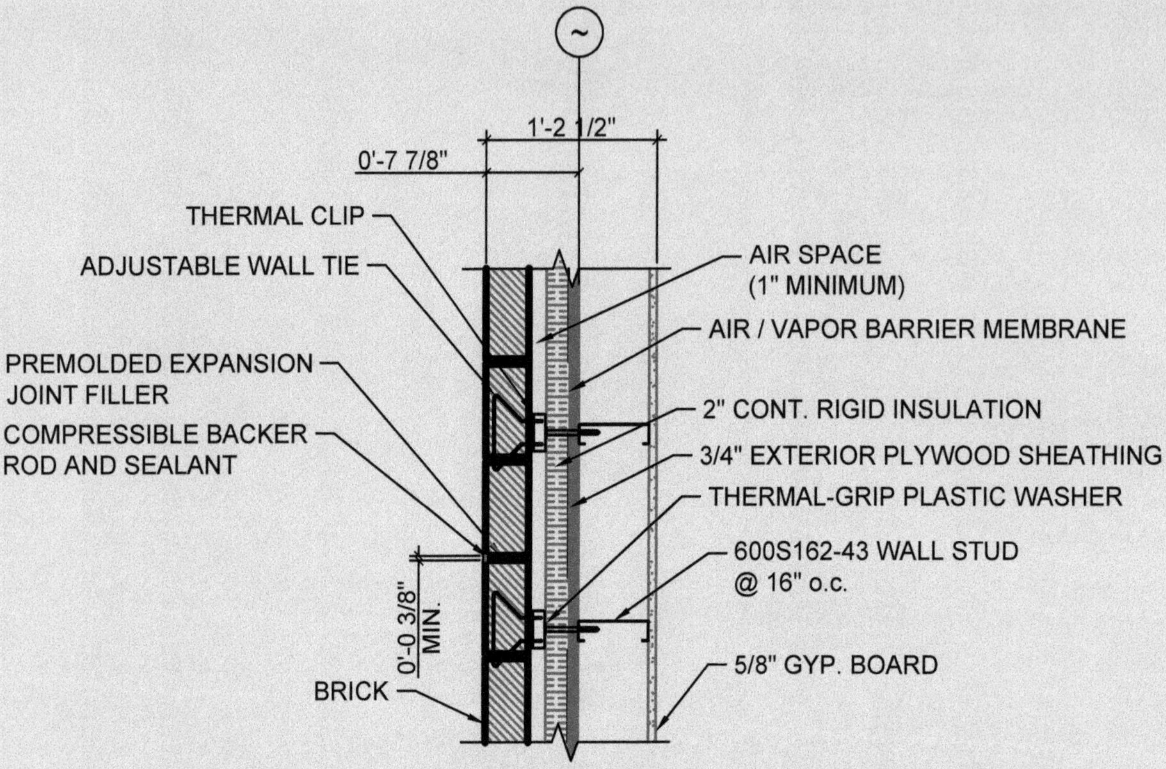

1 BRICK WALL MOVEMENT JOINT DETAIL

3/4" = 1'-0"

REFERENCE 9-11

BRICK WALL BACK FILLED WITH METAL STUDS WINDOW JAMB DETAIL

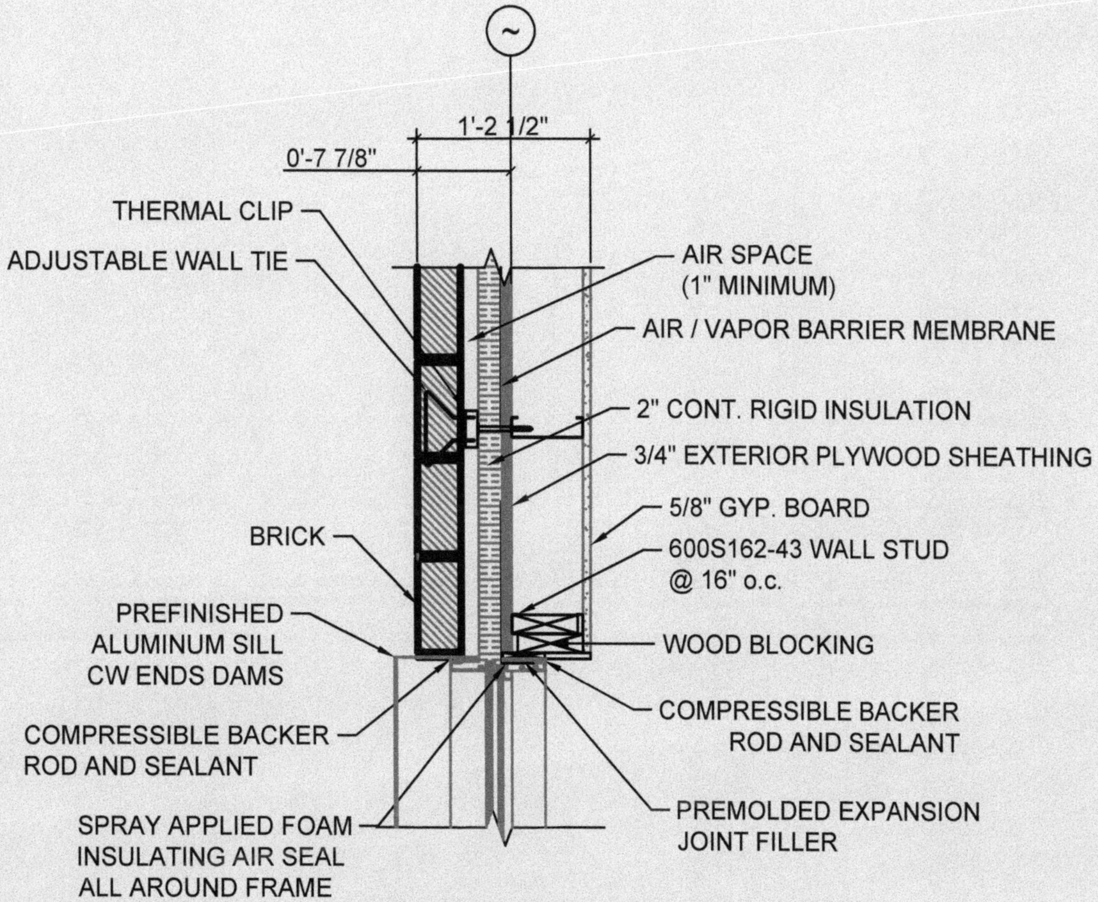

1 ALUMINUM WINDOW JAMB
(PLAN VIEW)
3/4" = 1'-0"

VENT FLASHING DETAIL AT ROOF

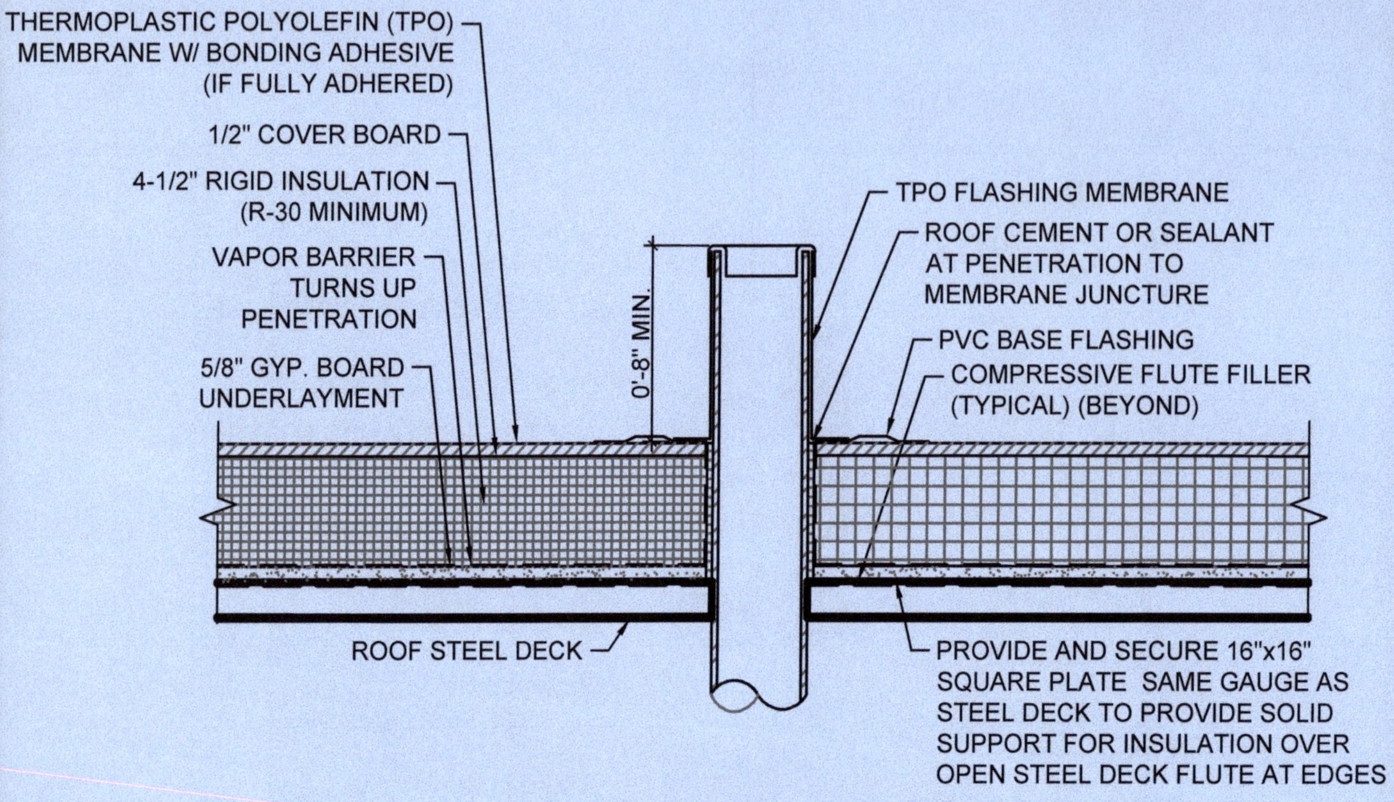

1 VENT FLASHING DETAIL
AT ROOF (TYPICAL) 1-1/2" = 1'-0"

TYPICAL ROOF DRAIN DETAIL

EXTEND TPO ROOFING MEMBRANE UNDER ROOF DRAINS CLAMPING RING AND SET ROOF DRAIN ONLY ON DRAIN RECEIVER AND SECURE TO ROOF CONSTRUCTION WITH DECK CLAMP ASSEMBLIES PER ROOFING MANUFACTURER AND ROOF DRAIN MANUFACTURER REQUIREMENTS.
SEE ARCHITECTURAL DRAWINGS FOR PLANS AND LOCATIONS OF DRAIN / SUM AREAS. SEE PLUMBING DRAWINGS AND SPECIFICATIONS FOR ROOF DRAIN MODEL NUMBERS AND SIZES.

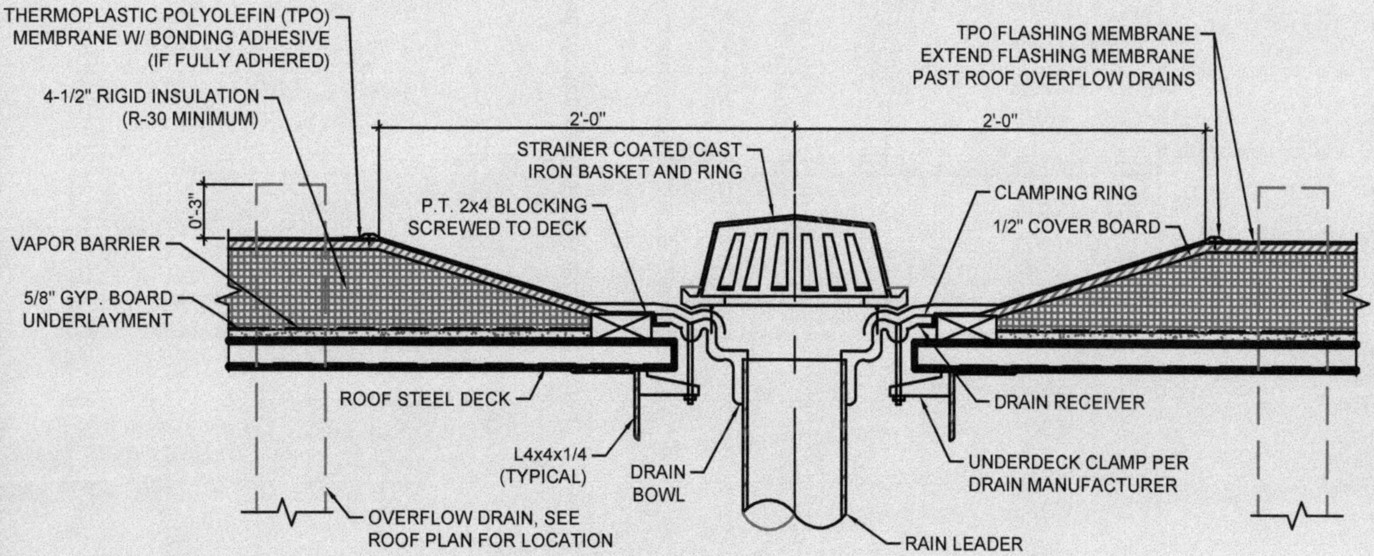

1 TYPICAL ROOF DRAIN DETAIL
AT ROOF (TYPICAL) 1-1/2" = 1'-0"

REFERENCE 9-14

PIPE FLASHING DETAIL AT ROOF

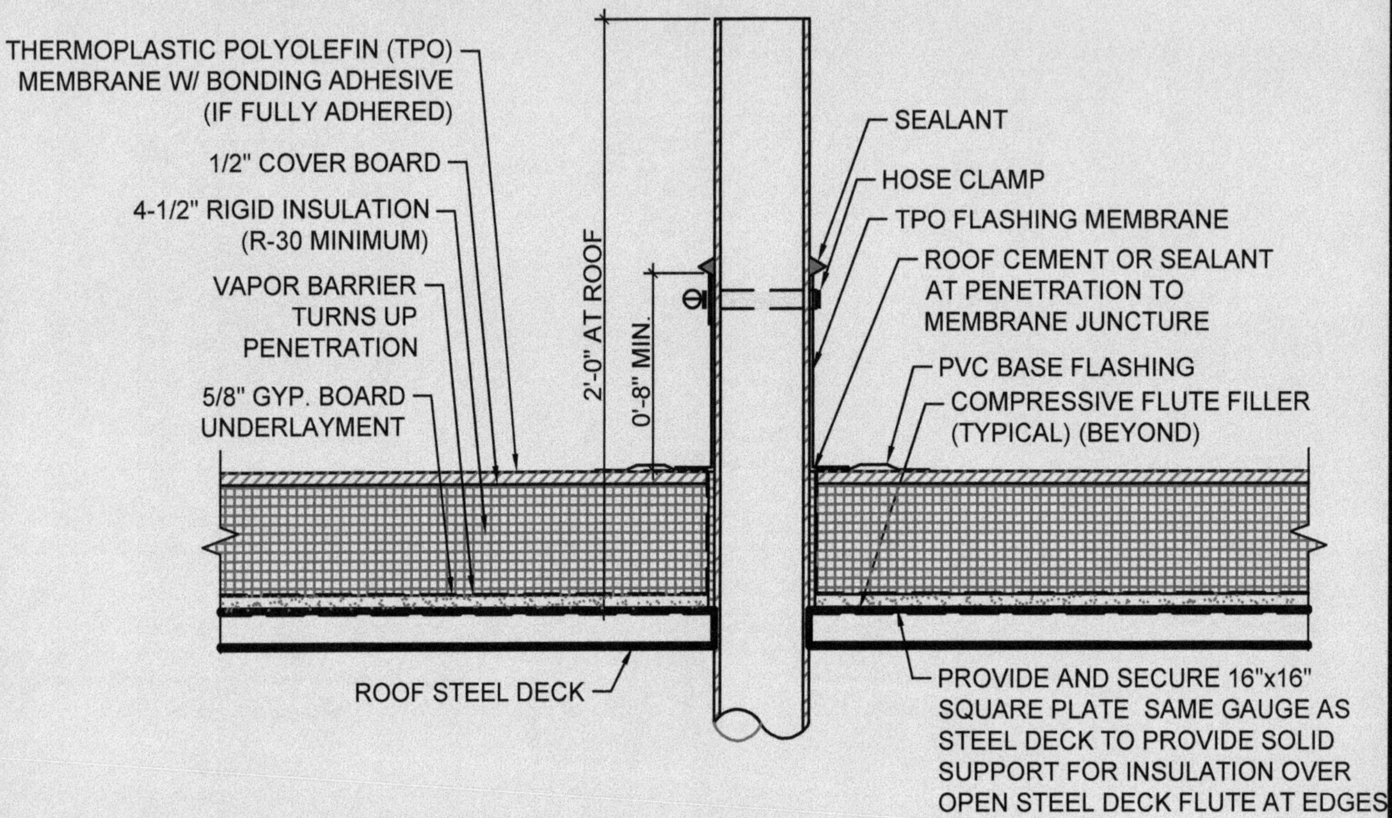

1 PIPE FLASHING DETAIL AT ROOF (TYPICAL) 1-1/2" = 1'-0"

REFERENCE 9-15

SCUPPER DETAIL AT ROOF

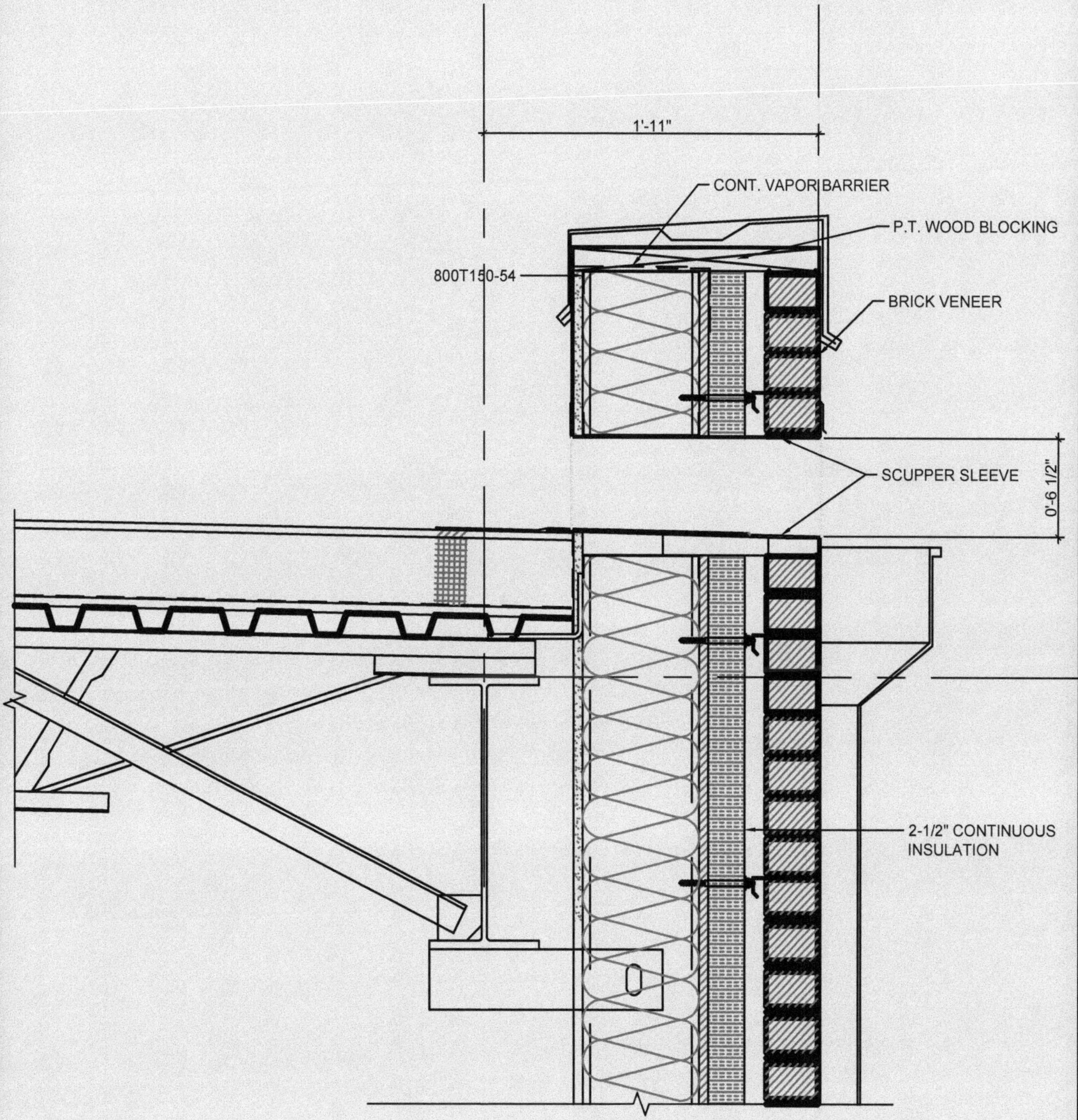

PROJECT NAME:	PROJECT #:
PROJECT ADDRESS:	DATE:
	INITIALS:

www.ingramcontent.com/pod-product-compliance
Lightning Source LLC
LaVergne TN
LVHW070854210926
848482LV00016B/933
* 9 7 9 8 3 4 8 4 4 2 8 1 1 *